Land drainage and flood defence responsibilities

Fourth edition

Institution of Civil Engineers

Land drainage and flood defence responsibilities

Fourth edition

Published by Thomas Telford Limited, 40 Marsh Wall, London E14 9TP.
www.thomastelford.com

Distributors for Thomas Telford books are
USA: ASCE Press, 1801 Alexander Bell Drive, Reston, VA 20191-4400
Australia: DA Books and Journals, 648 Whitehorse Road, Mitcham 3132, Victoria

First edition published 1983
Second edition published 1993
Third edition published 1996
Fourth edition published 2010

Also available from Thomas Telford Limited
Future flooding and coastal erosion risks. C. Thorne, E. Evans and E. Penning-Rowsell.
ISBN: 978-0-7277-3449-5
Planning and environmental protection: An introductory guide. Royal Town Planning Institute.
ISBN: 978-0-7277-3102-9
Design and access statements explained. Edited by R. Cowan, Urban Design Group.
ISBN: 978-0-7277-3440-2

www.icevirtuallibrary.com

A catalogue record for this book is available from the British Library

ISBN: 978-0-7277-3389-4

Institution of Civil Engineers ©2010

Typeset by Academic + Technical, Bristol
Index created by Indexing Specialists (UK) Ltd, Hove, East Sussex
Printed and bound in Great Britain by CPI Antony Rowe, Chippenham

Contents

Acknowledgement

The Institution of Civil Engineers is indebted to the revising author, Melinda Lutton, Civil Associate at Cundall, who has updated and rewritten large parts of the text for this fourth edition of *Land drainage and flood defence responsibilities*.

Preface

The third edition of *Land drainage and flood defence responsibilities* appeared in 1996, and was reprinted in 2001. Since this update, statutory bodies and legislation have changed significantly, making a revision long overdue.

The format of the new guide follows that of the third edition, but includes updates on a range of primary and secondary case law, such as the Civil Contingencies Act in 2004, the Water Act 2003 and a number of EU Directives such as the EU Floods Directive 200760/EC on the assessment and management of flood risks. In addition, environmental management and the planning process have changed significantly. Planning Policy Statement 25: Development and Flood Risk has meant that a risk-based approach is now taken for flood risk and management for all forms of flooding. Greater control of development in flood plains has resulted, with flood risk management now being considered as part of the planning process. Coastal erosion is seen as an integral part of managing flood risk.

It is likely that a further review will be required within the next year or two, following implementation of further new legislation. The draft Flood and Water Management Bill is under consultation, and a lot of work is being done to implement the recommendations of the independent review by Sir Michael Pitt following flooding in 2007. The Scottish Government is also consulting on a draft flooding Bill. In addition, the Agency has moved from a position of flood defence to flood risk management.

I would like to thank Greg Lutton (Parsons Brinckerhoff) and Ian Cansfield (Cundall), who edited chapters on environmental legislation and planning law. In addition, assistance with the text was given by David Balmforth (MWH) and Caroline McGahey (HR Wallingford). Thanks also to Ann Webster, who deciphered my handwriting and turned this into a first draft.

Finally, although every care has been taken in the revision of this publication, neither its authors nor their organisations can accept any legal liability for its contents, which do not necessarily represent the views of the sponsoring organisations.

Melinda Lutton
Civil Associate, Cundall

Abbreviations and definitions

Abbreviations

Agency	Environment Agency (for England and Wales)
CA	Coal Authority
CCA 2004	Civil Contingencies Act 2004
CCW	Countryside Council for Wales
CFMP	Catchment Flood Management Plans
CIA 1994	Coal Industry Act 1994
CMP	Catchment Management Planning
CMSA	Coal Mining Subsidence Act 1991
COW	Critical ordinary watercourse
CPA 1949	Coast Protection Act 1949
DCLG	Department of Communities and Local Government
Defra	Department of Environment Food and Rural Affairs
DoE	Department of Environment
EA 1995	Environment Act 1995
EIA	Environmental Impact Assessment
ES	Environmental Statement
EU	European Union
GDPO	Town and Country Planning (General Development Procedure) Order 1995
HA 1980	Highways Act 1980
IDB	Internal Drainage Board
LDA 1991	Land Drainage Act 1991
LDA 1994	Land Drainage Act 1994
LDD	Local Development Document
LFDC	Local Flood Defence Committee
MCAB	Marine and Coastal Access Bill 2008–9
NRA	National Rivers Authority
PCPA 2004	Planning and Compulsory Purchase Act 2004

PHA 1936	Public Health Act 1936
PPS25	Planning Policy Statement 25: *Development and Flood Risk*
RFDC	Regional Flood Defence Committee
RFRA	Regional Flood Risk Appraisal
S.	Section number of an Act of Parliament
SFRA	Strategic Flood Risk Assessment
SMP	Shoreline Management Plan
SS.	Section numbers of an Act of Parliament
SSSI	Site of Special Scientific Interest
SUDS	Sustainable drainage system
SWMP	Surface Water Management Plan
TCPA 1990	Town and Country Planning Act 1990
WA	National Assembly for Wales ('Welsh Assembly')
WA 2003	Water Act 2003
WAG	Welsh Assembly Government
WCA 1981	Wildlife and Countryside Act 1981
WIA 1991	Water Industry Act 1991
WRA 1991	Water Resources Act 1991

Statutory definitions

Critical ordinary watercourse (COW). Means a subdivision of an ordinary watercourse identified as most likely to flood properties. The Agency is taking over responsibility for COWs.

Drainage. Includes defence against water (including sea water), irrigation other than spray irrigation, warping and the carrying on, for any purpose, of any other practice which involves management of the level of the water in a watercourse (S.72(1) LDA 1991 and S.113(1) WRA 1991 as amended by S.100 EA 1995). References in the LDA 1991 to the carrying out of drainage works include references to the improvement of drainage works (S.72(5) LDA 1991).

Drainage body. Means the Agency, an internal drainage board or any other body having the power to make or maintain works for the drainage of land (S.72(1) LDA 1991).

Flood defence. Means the drainage of land and the provision of flood warning systems (S.113(1) WRA 1991).

Flood warning system. Means any system whereby, for the purpose of providing warning of any danger of flooding, information with respect to specified matters is obtained and transmitted whether automatically or otherwise, with or without provision for carrying out calculations based on such information and for transmitting the results of those calculations. The specified information is with respect to: (a) rainfall, as measured at a particular place within a particular period; (b) the level or flow of any inland water, or any part of an inland water, at a particular time; and (c) other matters appearing to the Agency to be relevant to providing warning of any danger of flooding.

Local authority. Means the council of a county, county borough, district or London borough or the Common Council of the City of London (S.72(1) LDA 1991, and S.221(1) WRA 1991).

Main river. Means a watercourse shown as such on a main river map and includes any structure or appliance for controlling or regulating the flow of water into, in or out of the channel which: (a) is a structure or appliance situated in the channel or in any part of the banks of the channel and (b) is not a structure or appliance vested in or controlled by an Internal Drainage Board (S.113(1) WRA 1991, and see S.137(4) WRA 1991).

Ordinary watercourse. Means a watercourse that does not form part of a main river (S.72(1) LDA 1991, and see the definitions of 'main river' and 'critical ordinary watercourse' above).

Public sewer. Means a sewer for the time being vested in a sewerage undertaker in its capacity as such, whether vested in that undertaker by virtue of a scheme under Schedule 2 to the WA 1989, S.179 of or Schedule 2 to the WIA 1991 or otherwise (S.221(1) WRA 1991, and S.219(1) WIA 1991).

Sewer. Includes all sewers and drains which are used for the drainage of buildings and yards appurtenant to buildings, excluding a drain used for the drainage of one building or of buildings or yards appurtenant to buildings within the same curtilage (S.219(1) WIA 1991 and S.221(1) WRA 1991). References to a 'sewer' are to include references to a tunnel or conduit which serves similarly or to any accessories thereof (S.219(2) WIA 1991 and S.221(2) WRA 1991).

Watercourse. Includes all rivers and streams and all ditches, drains, cuts, culverts, dikes, sluices, sewers (other than public sewers within the meaning of the WIA 1991) and passages through which water flows (S.72(1) LDA 1991, and see the definition of 'public sewer' above; similarly, see S.113(1) WRA 1991 but contrast S.221(1) WRA 1991).

1

Outline of responsibilities and statutes

Scope of land drainage

1. Land drainage and flood defence are generally understood to include the alleviation or control of flooding of urban and agricultural land, whether by fresh water or salt water, including the improvement and maintenance of natural and artificial channels used for these purposes.
2. Historically, land drainage was concerned particularly with the protection of arable land and the improvement of agricultural productivity by ensuring the optimum level of moisture in the soil. More recently, it has become increasingly directed by concerns that drainage activities should be conducted in accordance with environmental and conservation objectives.
3. In S.72 of the Land Drainage Act 1991 (LDA 1991), drainage was defined as including 'defence against water (including sea water), irrigation, other than spray irrigation, and warping'. An identical definition was given in S.113 of the Water Resources Act 1991 (WRA 1991) which also defines 'flood defence' as the 'drainage of land and the provision of flood warning systems'. The general emphasis on defence is somewhat typical of most land drainage legislation, but the definition has been extended by wording which makes clear that flood defence functions also include the management of water levels (S.100 of the Environment Act 1995 (EA 1995)).
4. Land drainage deals with natural flow, and therefore excludes both the drainage of water from artificial surfaces by means of pipes and culverts (i.e. surface water sewerage) and the protection of the coastline from erosion (i.e. where, unlike flooding, the coast is backed by high land). The Coast Protection Act 1949 (CPA 1949) deals with the problem of coastal erosion, soon to be updated by the MCAB. The WRA 1991 and LDA 1991 deal with flooding, updated by the WA 2003.

General principles

5. Land drainage activities provide no absolute guarantee against flooding. The level of protection provided will depend in each case on the extent and cost of the measures undertaken. While other service industries may strive to supply the full requirements of consumers, land drainage work is generally limited to the degree of protection agreed between those concerned.

6. Legislation dealing with land drainage has existed in England and Wales for at least five and a half centuries. There is even evidence that the Romans set up quite complex organisations to deal with land drainage. The statutes and common law have developed out of the need to resolve practical problems, and to provide a workable allocation of responsibilities, so that the law embodies lessons from longstanding practical experience.

7. Land drainage legislation was uncertain and fragmented prior to the important Land Drainage Act 1930, which consolidated and greatly clarified it. Further consolidations were effected by the Land Drainage Acts of 1976 and 1991, the latter following the changes (including establishment of the National Rivers Authority (NRA)) made by the Water Act 1989, further amended by the WA 2003. The LDA 1991 re-enacts most of the previous land drainage provisions, but those relating to the Environment Agency (see below) and main rivers appear in the WRA 1991. The Land Drainage Act 1994 (LDA 1994) adds new environmental duties to the LDA 1991.

8. Under the EA 1995, provision was made for the establishment of an Environment Agency (the 'Agency') for England and Wales to which the functions of the NRA were transferred on 1 April 1996. The Agency also took over the functions of Her Majesty's Inspectorate of Pollution and the waste regulation functions of local authorities. The main powers and duties of the Agency are detailed in Chapter 3 below.

9. A general distinction is to be made between a main river and an ordinary watercourse. These terms are defined at the front of this book in the section on statutory definitions. Under the WRA 1991, the Agency has responsibilities for main rivers. These are exercised largely through Regional Flood Defence Committees (RFDCs). Local authorities and Internal Drainage Boards (IDBs) have responsibilities for all other watercourses, termed ordinary watercourses, largely under the LDA 1991. Critical ordinary watercourses (COWs) are a subdivision of ordinary watercourses, and these were identified by the Government following widespread flooding in 1998. They are the

watercourses identified as those most likely to flood the equivalent of 25 properties in any 1 kilometre stretch. In 2003, the Department of Environment Food and Rural Affairs (Defra) decided that all COWs should be designated as main rivers.

10.   Common law precedents and statutory provisions have established the general principles which govern the present arrangements, and these are summarised below:

(a) Individual owners are responsible for the drainage of their own land, and for accepting and dealing with the natural catchment flows from adjoining land. They must not permit an obstruction to the natural flow without consent.

(b) Powers given to public authorities are, in general, permissive, thereby recognising the rights and obligations of riparian owners and other individuals, and giving such authorities a degree of discretion over public expenditure priorities.

(c) Permissive powers are available to a local authority to enable it to carry out flood protection works on ordinary watercourses. Such work may therefore be funded directly from the rates and charges within the local area concerned.

(d) Defra grant aid and other direct contributions are additional sources of funds, but the underlying principle is that, except in the case of main rivers, the decision to give priority to funding to carry out works on a particular watercourse rests with the individual riparian owners or with the local authority, i.e. those who derive local benefit from the works.

(e) In the case of designated main rivers which carry water from the upland areas through to the sea, certain powers and duties rest with the Agency, which draws its funds from the whole catchment. Apart from the mandatory duty of general supervision, the statutory provisions concerning main rivers confer discretionary powers of control and powers to do work, including maintenance and improvements.

(f) Generally, riparian ownership and landowners' responsibilities on such rivers remain unaffected, together with the associated common law rights and obligations.

(g) S.33 LDA 1991 (with S.107 WRA 1991) provides, with respect to main rivers, that the Agency shall 'take steps to commute' certain obligations by way of a sum assessed under S.34 LDA 1991. The 'obligations' are those imposed on persons 'by reason of tenure, custom, prescription or otherwise to do any work in connection with the drainage of land (whether by way of repairing banks or walls, maintaining watercourses or otherwise)', so that the application of the section is limited.

(h) Where similar obligations apply to ordinary watercourses, the Agency and IDBs may, with the consent of the Secretary of State, exercise their discretion to commute these obligations (see S.33(l) LDA 1991). The sums to be paid are determined in accordance with the provisions of S.34 LDA 1991.

Statutory functions

11.   Apart from the principal Acts (the WRA 1991, LDA 1991 as amended by the LDA 1994, and EA 1995) there are many other statutes, particularly those dealing with environmental, local government and public health matters. A list of the main enactments and Government policy guidance is given in Appendix 1 below.

12.   A brief outline of the main provisions of the WRA 1991 and LDA 1991 is given below. Details of the relevant provisions of other Acts, Circulars, etc., are given in the relevant sections of this guide.

13.   Under the provisions of the WA 1989 (see now WRA 1991) the relevant powers and duties of the former water authorities were transferred to the NRA. However, the statutory powers and duties of land drainage authorities and bodies were not fundamentally changed. The EA 1995 transferred the functions of the NRA to the Agency but did not fundamentally alter the relevant statutory powers and duties, which remain in the WRA 1991 and LDA 1991, though the expanded definition of drainage to encompass water level management should be noted.

14.   S.6(4) EA 1995 reaffirms the supervisory role of the Agency throughout England and Wales over all matters relating to flood defence.

15.   It should be noted that the Water Act 1973 and later legislation did not alter the previous important distinctions between sewerage and land drainage functions. What was a land drainage function before the 1973 Act remained a land drainage function under the WRA 1991 and LDA 1991. Similarly, what was a sewerage function remains a sewerage function under the Water Industry Act 1991 (WIA 1991), subsequently amended by the WA 2003.

16.   For current law, reference should largely be made to the WRA 1991 and LDA 1991 as amended. These Acts laid down no new duties on public authorities as they were consolidation Acts. The powers which they confer on drainage bodies therefore remain unchanged in most respects, and are mainly permissive. However, most of the general duties placed on the Agency and Secretaries of State are now contained in the EA 1995.

17.　　The system of land drainage in England and Wales provided for by the WRA 1991 and LDA 1991 may be regarded as a 'tiered system' comprising the following authorities and bodies:

(a) *Central government.* Defra and the National Assembly for Wales ('Welsh Assembly' – WA) have overall policy responsibility for land drainage and flood defence matters. The Agency and the WA have a key role in relation to authorising works for which planning permission may be required.

(b) *The Environment Agency.* The Agency must exercise 'a general supervision' over all matters relating to flood defence (S.6(4) EA 1995), but it must delegate all its own land drainage functions except levies, charges and borrowing to RFDCs (S.106WRA 1991).

(c) *Regional Flood Defence Committees.* RFDCs comprise members appointed by central government, the Agency and relevant local authorities (S.15 EA 1995). They may, in turn, delegate functions to Local Flood Defence Committees (LFDCs), which are usually based on the districts of former river authorities. LFDCs are made up of appointees from the RFDC and from relevant councils (S.18 EA 1995). Within the districts, IDBs and internal drainage districts may continue to exist in some areas.

(d) *Internal Drainage Boards.* IDBs were originally set up to deal with specific problems in identifiable areas, and they are the 'drainage body' for Internal Drainage Districts. As such, they are under a duty to exercise general supervision over all matters relating to the drainage of land within their district (S.1(2) LDA 1991).

(e) *Internal Drainage Districts.* These are statutorily defined as 'such areas within the areas of regional flood defence committees as will derive benefit or avoid danger as a result of drainage operation' (S.1 LDA 1991). The above arrangements allow a fair degree of localised control of budgeting and decision-making within the framework of the Agency's supervisory role.

(f) There are also local authorities and, uncommonly, local commissioners, conservators, etc., or other bodies, which may have inherited specific local powers to make or maintain works for the drainage of land.

18.　　In general, local authorities may exercise certain statutory powers in respect of ordinary watercourses, which are not (a) main rivers, which are subject to the direct discretionary powers given to the Agency, or (b) subject to similar control powers given to IDBs. Since the powers of an IDB enable it to deal with all watercourses within its district, they overlap with the similar powers of local authorities (see 4.1–4.6 below).

19.   Other private organisations or public bodies may exercise comparatively minor drainage powers, or carry out drainage activities because of common law obligations or local Acts. Examples are the Coal Authority and licensed coal operators, British Waterways, Highways Agency and highway authorities. The last are responsible for draining highways, which includes preventing water from flowing onto a highway, and have responsibility for certain bridges and culverts, under the Highways Act 1980 (HA 1980) (see Chapter 12 below).

20.   Navigation authorities have also been given certain powers and responsibilities for drainage; these are usually incorporated in special Acts which authorise the construction of the navigation works. These specialised responsibilities are often mandatory, incorporating a requirement to make good any drainage deficiency which is caused by the works.

21.   The draft Flood and Water Management Bill was published in consultation in April 2009. This is intended to introduce measures to protect the UK from the possible effects of climate change and to reduce the risk of disruption to households, businesses and the economy caused by flooding and drought.

Principal statutes

22.   Under the EA 1995, LDA 1991 and WRA 1991, the Secretary of State for Environment, Food and Rural Affairs has various statutory duties and powers in relation to the Agency, IDBs and also local authorities where they act as drainage bodies.

Environment Act 1995

23.   The EA 1995 provides the general framework for the Agency, its constitution and general duties. Provisions of particular importance concerning the role of Defra include the following:

- *General oversight of the Agency.* The Secretary of State appoints three of the members of the Agency and makes appointments (including that of the Chairman) to RFDCs. The Secretary of State may give the Agency specific or general directions: for example, as to how the Agency is to contribute to achieving sustainable development (S.4(3) EA 1995).

- *Enhancement of the environment.* In respect of land drainage and flood defence functions, the EA 1995 places a duty on the Secretary of State and the Agency to further conservation and the conservation and enhancement of natural beauty so far as may be consistent with enactments relating to their functions and objectives or guidance relating to the achievement of sustainable development (see further in Chapter 17 below).

Water Resources Act 1991

25. The WRA 1991 contains legislation relating to the Agency, including flood defence legislation previously in the Land Drainage Act 1976. Provisions concerning the role of Defra of particular importance to engineers include the following:

- *Designation of a main river.* The definitive map is issued by Defra. The Agency may propose changes for confirmation by the Secretary of State (S.194 WRA 1991).
- *Grants.* SS.147–149 WRA 1991 enabled the relevant Secretary of State to make grants to the Agency for drainage works, flood warning schemes, bridge works and ancillary items, including land purchase and compensation (see also Chapter 14 below). The WA 2003 has amended the WIA 1991 and WRA 1991, and included various improvements to streamline arrangements for flood defence organisation and funding, repealing sections 147 to 149 of the WRA 1991.
- *Land and works powers.* Compulsory purchases of land by the Agency must be authorised by the relevant Secretary of State, who also controls acquisition of land by accretion and the disposal of compulsorily acquired land (SS.154, 155 and 157 WRA 1991).

Land Drainage Act 1991

26. The LDA 1991 brought together legislation relating to IDBs and local authorities previously in the Land Drainage Act 1976, concerning inland and sea defence matters. As amended by the LDA 1994, provisions of particular importance to engineers include the following:

- *Organisation of IDBs.* Schemes for the constitution of IDBs must be confirmed by the Secretary of State. Operationally, IDBs are supervised by the Agency, but the Secretary of State determines matters in case of a dispute or where the Agency wishes to exercise default powers. The Secretary of State supervises the exercise of default powers by local authorities.

- *Enhancement of the environment.* The duty placed on the Secretary of State, the Agency and IDBs is similar to that in the EA 1995 (see 2.9 above, second provision).
- *Restoration and improvement of ditches.* If a ditch is in such a condition as to cause injury to land or to prevent improvement or where the drainage of land requires improvement, the Agricultural Land Tribunal may order remedial or improvement works which the Secretary of State has power to enforce. Defra provides a secretariat and expert support to Agricultural Land Tribunals, which are independent bodies. Contact can be made through Defra local offices or the Defra Regional Engineer (see Appendix 3 below).
- *Borrowing and contributions from the Agency.* The Secretary of State must give consent to any borrowing by IDBs. Where there is a dispute with the Agency about contributions to a Board, or allocation of Agency revenue if it is acting as a Board, the matter goes to the Secretary of State for resolution. (For contributions to the Agency, see S.139 WRA 1991.)
- *Grants to drainage bodies.* S.59 LDA 1991 enables Secretaries of State to make grants to IDBs or other drainage bodies for drainage schemes (see further in Chapter 14 below).
- *Compulsory purchase and disposal of land.* Where necessary, the Secretary of State may authorise compulsory purchase of land, and the Secretary of State also controls disposal of compulsorily purchased land (S.62(l)(b) LDA 1991).

Civil Contingencies Act 2004

27.   The CCA 2004 repeals the Civil Defence Act 1948 and defines 'emergency', which includes flooding. The Act imposes a series of duties on local bodies, including the Agency, to assess the risk of an emergency and to maintain plans for responding to it.

The EU Floods Directive 2007

28.   Directive 2007/60/EC on the assessment and management of flood risks requires Member States to assess if all water courses and coastlines are at risk from flooding, to map the flood extent and assets and people at risk, and to take adequate and coordinated measures to reduce the flood risk. The recommendations of the Directive shall be carried out in coordination with the WFD by flood risk management plans and river basin management plans. Defra is undergoing a consultation

process to transpose the Directive into UK law, with the draft Floods and Water Management Bill.

Marine and Coastal Access Bill 2008 – 9

29.   This Bill is in progress through parliament and will improve and simplify arrangements for managing marine development. This includes the introduction of a new marine planning system for the strategic management of the marine area around the UK. It aims to consolidate and modernise Part II of the CPA 1949, including replacement of licensing and consents.

Draft Floods and Water Management Bill 2009

30.   Defra and the WA published draft bill on 21 April 2009. This includes proposals relating to funding and investment, reforms to IDBs, whether developers are contributing adequately to flood risk management. The first reading of this Bill is due in May 2010.

2

Government departments

Department for Environment, Food and Rural Affairs (Defra)

Introduction

1.   Defra is the Government department with the overall policy responsibility for flood defence and coastal protection in England. One of the department's principal aims, by assisting in and encouraging the building of defences, is to prevent flooding and coastal erosion. Defra was formed in 2001 from part of the former Department for the Environment, Transport and the Regions (DETR), the Ministry of Agriculture, Fisheries and Food (MAFF) and a small part of the Home Office. The dissolution order transferred all property, rights, liabilities, statutory powers and duties of the MAFF to the Secretary of State for Environment, Food and Rural Affairs.

2.   Defra has national policy responsibility for flood and coastal erosion risk management, but does not manage or build flood defences, nor direct the authorities on which specific projects to undertake.

Policy

3.   *Surface water drainage.* Following flooding in the summer of 2007, the Government commissioned Sir Michael Pitt to conduct an independent review to assess what happened and what might be done differently in the future. The essential recommendations were:

- improve the quality of flood warnings and flood modelling
- a wider brief for the Agency and to ask local authorities to strengthen their capability to take the lead on flood risk management
- more robust building and planning controls to protect communities
- better preparation and information for emergency services
- better planning and higher levels of protection for critical infrastructure
- raise awareness through education and publicity programmes

- speed up the recovery time
- learn from international experience.

4. *Surface Water Management Plans.* Defra prepared a 'living draft' document, *Surface Water Management Plan Technical Guidance* in February 2009 to inform local authorities on how to approach the development of Surface Water Management Plans (SWMPs). These plans will provide the coordinated basis for managing local flood risk, and six pilot areas are being tried. This work is being overseen by a steering group including Defra, the Department for Communities and Local Government, and the Agency.

5. *Adaption and resilience.* Defra policy aims to reduce the impact of flooding and coastal erosion and to adapt to the effects of climate change. Defra is supporting the development of an 'adaption toolkit', which will focus on ways to reduce the damage caused by floods and coastal erosion, and comprises a strengthened planning policy for flood risk areas. Defra has prepared a coastal change policy, launched a property-level protection grant scheme and has recently funded a grant to look at the links between land management and flooding.

6. *Emergency planning.* Defra, the Agency and the Civil Contingencies Secretariat have produced an outline New National Flood Emergency Framework, which was under consultation until the end of 2008.

7. *Draft Flood and Water Management Bill.* This will create a simpler and more effective means of managing the risk of flood and coastal erosion. The draft bill is anticipated in autumn 2010 and aims to:

- deliver improved security, service and sustainability
- clarify responsibilities for flood risk
- protect essential water supplies by enabling water companies to control non-essential water use during drought
- modernise the law for managing the safety of reservoirs
- encourage sustainable drainage
- make it easier to resolve misconnections to sewers.

Roles and responsibilities of parties

8. Defra has overall policy responsibility for flood and coastal erosion risk in England. It funds most of the Agency's activities and provides grant aid to other flood and coastal defence operating authorities (local authorities and IDBs), to support their investment in improvement works. Improvement projects must meet specified economic, technical and environmental criteria to be eligible for funding.

9.   Communities and Local Government is a body responsible for spatial planning policy and the operation of the planning system in England. Design and flood resilience issues not related to external appearance are matters for the Building Regulations, also administered by Communities and Local Government.

10.   Government offices have a role to scrutinise draft regional strategies and Local Development Documents (LDDs), which will include flood risk policies. Where a local planning authority is minded to approve a planning application and there is Agency objection to it on flood risk grounds, the application must be referred to the appropriate Government office to consider, on behalf of the Secretary of State, whether it should be called in for determination.

11.   Local highways authorities have responsibility for ensuring that road drains are maintained. The Highways Agency is responsible for managing road drainage from the trunk road network in England, including the slip roads to and from trunk roads.

12.   Sewerage undertakers are generally responsible for surface water drainage from development via adopted sewers and, in some instances, sustainable drainage systems (SUDSs). They should ensure that Urban Drainage Plans reflect the appropriate Regional Spatial Strategies and LDDs, in line with their obligations in the current legislation.

13.   Certain reservoir undertakers will be required to produce emergency contingency plans (Flood Plans), following direction by the Secretary of State under the Reservoirs Act 1975, as amended. The presence of reservoirs and implications for flood risk should be recognised in Regional Flood Risk Appraisals (RFRAs), Strategic Flood Risk Assessments (SFRAs) and Flood Risk Assessments.

14.   The Civil Contingencies Act 2004 and associated Regulations set out an emergency preparedness framework, including planning for and the response to emergencies. Local Resilience Forums, which include representatives from the emergency services, local authorities and the Agency, should ensure that risks from flooding are fully considered, including the resilience of the emergency infrastructure that will have to operate during floods. Emergency services should be consulted during the preparation of LDDs and the consideration of planning applications where emergency evacuation requirements are an issue.

15.   The Agency was established by the Environment Act 1995, and is a non-departmental public body of Defra. It is the principal flood defence operating authority in England. Under the WRA 1991, the Agency has permissive powers for the management of flood risk arising from designated main rivers and the sea. The Agency is also responsible for flood

forecasting and flood warning dissemination, and for exercising a general supervision over matters relating to flood defence. The Agency is required to arrange for all its flood defence functions (except certain financial ones) to be carried out by RFDCs under S.106 WRA 1991. It also assesses applications for funding of flood management schemes.

16.　　Local authorities have certain permissive powers to undertake flood defence works under the LDA 1991 on watercourses which have not been designated as main rivers and which are not within IDB areas. There are also over 80 maritime district councils which have powers to protect the land against coastal erosion under the CPA 1949, soon to be amended by the MCAB. Local authorities can control the culverting of watercourses under S.263 of the Public Health Act 1936 (PHA 1936).

17.　　IDBs are independent bodies, created under various statutes to manage land drainage in areas of special drainage need. Each Board operates within a defined area in which it has permissive powers under the LDA 1991 to undertake flood defence works, other than on watercourses that have been designated as 'main'.

Grant aid

18.　　Chapter 14 below describes arrangements for grant aid, including information about grant aid from the WA.

Advisory function

19.　　Defra offers two sources of advice on flood defence and land drainage matters. Along with ADAS, engineers of the Flood Management Division's River and Coastal Engineering Group are able, subject to availability, to offer advice on all aspects of flood defence improvement schemes, including technical, environmental and economic aspects.

National Assembly for Wales (WA)

20.　　The WA covers the functions of most of the other government departments for Wales, including flood defence. As such, it undertakes Defra functions under the WRA 1991 and the LDA 1991 in the principality or, where appropriate, in relation to the Welsh region of the Agency.

21.　　The Secretary of State for Wales has powers concurrently with the Secretary of State for Environment, Food and Rural Affairs; in practice the Welsh Secretary appoints one member of the Agency and makes the appointments to the Welsh RFDC (SS.15 and 1 EA 1995). The Welsh Secretary must also establish and maintain a committee of his appointees to advise him about the work of the Agency in Wales (S.11 EA 1995).

22.   Within the WA, flood defence and coastal protection are administered by a branch of the Environment Division, which also contains the necessary expertise to provide technical advice.

23.   WA and Defra flood defence staff liaise closely, and there is generally little difference in policy and approach between the two Departments. Defra's *Making Space for Water* and the *Flood and Coastal Defence Project Appraisal Guidance Notes* are published jointly between Defra and the WA, and the WA was also consulted on the *Environment Manuals*.

Scottish Government

24.   The Scottish Government similarly covers the functions for flood defences in Scotland. This was established in 1999 as the Scottish Executive, and rebranded following the 2007 Scottish Parliament election. The structure of the ministerial team includes the Cabinet Secretary for Rural Affairs and the Environment and a Minister for Environment. There is considerable flux in this area at present. The EU Floods Directive being implemented through primary legislation is likely to give the Scottish Environment Protection Agency the strategic responsibility for flood risk management in Scotland.

3

Environment Agency

Supervisory duty

1. Under S.6(4) EA 1995, a duty is imposed on the Agency in relation to England and Wales to 'exercise a general supervision over all matters relating to flood defence'. A similar duty was placed on the NRA and, previously, on catchment boards in earlier provisions going back to the Land Drainage Act 1930. Supervision is one of the few mandatory duties imposed on public authorities in land drainage legislation.

2. Certain watercourses may be designated as 'main rivers' on a statutory map by the Secretary of State for Environment, Food and Rural Affairs (SS.193 and 194 WRA 1991), and the supervisory duty is usually exercised in some detail.

3. Nevertheless, the act of 'maining' a particular watercourse does not place any specific obligations on the Agency to exercise its permissive powers. The basic common law obligations of riparian owners therefore remain, as do, perhaps more importantly, their statutory responsibilities.

4. In this connection, S.107(2) WRA 1991 and S.21 LDA 1991 enable the Agency and IDBs to enforce certain obligations to which landholders are subject by reason of tenure, custom and prescription (but see also 3.32 below and 1.10(g) above).

5. S.107 WRA 1991 defines the permissive powers which are available to the Agency in respect of main rivers, and deals with maintenance, improvement works and construction. S.14(2) LDA 1991 allocates to the Agency powers in respect of main rivers, although, under S.11 LDA 1991, the Agency may agree to work being done on a main river by an IDB.

6. The Agency is not empowered to compel a local authority to carry out works on ordinary watercourses, its powers in this regard being limited to specific and limited 'consenting' powers. S.16 LDA 1991, however, enables powers not being used by a district council in England to be exercised by the county council; in the case of a metropolitan district

council, London borough council, Welsh county council or county borough council or the Common Council of the City of London (referred to in this guide generally as 'unitary authorities': see further in Chapter 5 below), it is the Agency which may act on request or after giving notice.

7.   It must therefore fall to others, e.g. local authorities and riparian owners, to exercise their powers and to assume their responsibilities in relation to ordinary watercourses in a responsible way.

8.   The Agency's supervisory duty is also exercised through its planning, consultation and liaison arrangements with local authorities. Planning Policy Statement 25 (PPS25) and Article 10 of the Town and Country Planning (General Development Procedure) Order 1995 (GDPO) specify the categories of development in relation to which consultation is required. In particular, planning authorities are directed to allow the Agency the opportunity to comment on a range of planning applications where, among other things, the development is proposed for a site within the flood plain or washland or within a coastal flood plain; within or adjacent to any watercourse; where the increased volume of run-off would be significant; which is likely to involve culverting or the diversion of a watercourse, where the development area exceeds 1 hectare; or where flooding from any other source is considered to be a risk.

9.   A further aspect of the supervisory role is exemplified by S.17 LDA 1991, which requires local authorities to consult with the Agency before exercising any of the general drainage powers conferred upon councils by SS.14–16 LDA 1991. Where the works are within an internal drainage district, the Agency must consult the IDB.

10.   When the powers conferred by S.18 LDA 1991, concerning schemes for the drainage of small areas, are exercised by unitary authorities, the Agency must be consulted in accordance with the requirements of S.18 and Schedule 4, paragraph 1, of the LDA 1991.

11.   Under the terms of S.339 HA 1980, a highway authority is required to obtain consent from the Agency or drainage body before works are carried out on any watercourse or drainage channel.

12.   Under S.266 PHA 1936, local authorities must consult the Agency or the relevant IDB in relation to SS.259–265 PHA 1936 (see Chapter 5 below).

Surveying duty

13.   The Agency has a key role to play in carrying out surveys of those areas in relation to which it carries out its flood defence functions, primarily main rivers and sea defence works (S.105(2) WRA 1991).

14.　The results of these surveys should be copied, as soon as they become available, to local planning authorities, in order to inform their development planning and development control functions.

15.　The role of the Agency generally in relation to planning matters is considered in more detail in Chapter 10 below.

Principal powers — main rivers and ordinary watercourses

16.　For the practising Agency engineer dealing with main rivers, SS.165, 107 and 109 WRA 1991 and S.23 LDA 1991 are probably the most important provisions, and are worth examining in detail here. Basically, these sections give the Agency comprehensive control powers over the main river elements of the system, but with lesser control over ordinary watercourses.

17.　S.165(1) WRA 1991 and S.14 LDA 1991 make it clear that powers over main rivers are to be exercised by the Agency, while other drainage bodies deal with ordinary watercourses.

18.　S.165(1) WRA 1991 states that the Agency has the power, in connection with a main river:

(a) to maintain existing works, i.e. to cleanse, repair or otherwise maintain in a due state of efficiency any existing watercourse or any drainage work

(b) to improve any existing works, i.e. to deepen, widen, straighten or otherwise improve any existing watercourse or to remove or alter mill dams, weirs or other obstructions to watercourses, or to raise, widen or otherwise improve any existing drainage work

(c) to construct new works, i.e. to make any new watercourse or drainage work or to erect any machinery or do any other act (other than an act referred to in paragraph (a) or (b) above) required for the drainage of any land.

19.　Under S.165(2) WRA 1991, the Agency also has the power, irrespective of whether or not the works are in connection with a main river, to maintain, improve or construct drainage works for the purpose of defence against sea water or tidal water. This power may be exercised both above and below the low-water mark. Section 165(3) authorises the Agency to construct all such works and do all such things in the sea or in any estuary as may, in its opinion, be necessary to secure an adequate outfall for a main river.

20.   S.23 LDA 1991 and S.109 WRA 1991 are also important to the engineer. It is noteworthy that the powers contained in S.23 LDA 1991 are available in respect of ordinary watercourses, while S.109 WRA 1991 applies only to main rivers.

21.   S.23(1) LDA 1991 provides that no person may:

(a) erect any mill dam, weir or other like obstruction to the flow of any ordinary watercourse or raise or otherwise alter any such obstruction or

(b) erect any culvert that would be likely to affect the flow of any ordinary watercourse or alter any culvert in a manner that would be likely to affect any such flow, without the consent in writing of the drainage board concerned.

22.   Consent must not be unreasonably withheld by the drainage board. Such consents may be given by the Agency unless there is an IDB in the particular area.

23.   S.109 WRA 1991 applies only to main rivers, and specifies that:

(1) no person shall erect any structure in, over or under a watercourse which is part of a main river except with the consent of and in accordance with plans and sections approved by the Agency

(2) no person shall, without the consent of the Agency, carry out any work of alteration or repair on any structure in, over or under a watercourse which is part of a main river if the work is likely to affect the flow of water in the watercourse or to impede any drainage work

(3) no person shall erect or alter any structure designed to contain or divert the floodwater of any part of a main river except with the consent of and in accordance with plans and sections approved by the Agency (S.109(l)–(3) WRA 1991).

24.   A consent may not be unreasonably withheld (S.110 WRA 1991); however, the Agency can make a charge for the issuing of a consent. The Agency has the power to remove unauthorised work and to recover the costs of doing so from a person who contravenes the above provisions (S.109(4) WRA 1991).

25.   SS.109(1) and (2) WRA 1991 do not apply to works carried out in an emergency, but where such a situation arises, the Agency must be notified as soon as practicable (S.109(5) WRA 1991).

26.   It follows that if a drainage authority spends money on the improvement of a watercourse, in exercising the powers provided by S.165 WRA 1991 or S.14 LDA 1991, there should be powers to protect such works

thereafter. Safeguards are provided by way of consent procedures to prevent subsequent problems from arising.

Designation of main rivers

27.   There are no statutory criteria (or non-statutory guidelines) for the designation of main rivers, other than being those shown on a main river map, and can include any structure or appliance for controlling the water in or out of a main river. The Agency is not required to justify an application to a Secretary of State for a particular watercourse to be designated, although there is an objection procedure under S.194 WRA 1991. Designation of a main river may be brought about in order to improve its standard, but, alternatively, it might be done as a precaution to ensure that the Agency has sufficient powers (e.g. to remove fallen trees) should a need arise for these powers to be exercised (see further in Chapter 9 below).

28.   There are about 21 000 miles of main rivers in England and Wales, ranging from the major rivers to smaller streams which drain only a few hundred hectares. Many of these watercourses receive little or no maintenance or improvement, and may not even be subject to regular inspection. The application of byelaws, however, considerably assists the process of control, and will often be sufficient to ensure that reasonable care and maintenance are exercised by riparian owners and other users of the river.

Powers to ensure maintenance of flow in watercourses

29.   S.25(1) LDA 1991 provides that where any ordinary watercourse is in such a condition that the proper flow of water is impeded, then (unless the problem is due to mining operations) the drainage board or local authority concerned may serve a notice to require the condition to be remedied. In default, the drainage board or local authority may execute the work itself and recover the expenses from the person on whom the notice was served (S.25(6) LDA 1991). S.25(4) LDA 1991 provides that serving notice on a person other than the owner is subject to the consent of the owner and occupier. S.27 LDA 1991 provides for a right of appeal to a magistrates' court.

30.   The drainage board or local authority concerned is defined in S.25(2) LDA 1991, but under S.107(3) WRA 1991 these powers are similarly exercisable by the Agency in relation to main rivers. In relation to ordinary watercourses, the powers are exercisable:

(a) in an internal drainage district, by the IDB and the local authority, or
(b) outside an internal drainage district, by the Agency and the local authority.

The powers in each case are subject to the provisions of S.26 LDA 1991, as described below.

31.   Under S.26 LDA 1991, before exercising powers under S.25 LDA 1991, a local authority must notify the IDB (in an internal drainage district) or the Agency. Where a local authority has relevant powers other than those under S.25 LDA 1991, a drainage body may not exercise the powers under S.25 except by agreement with the local authority, or in default after reasonable notice. Where there is a navigation, harbour or conservancy authority, the consent of that authority is required. S.26 LDA 1991 does not apply to main rivers.

Powers to require repairs to watercourses, bridges, etc.

32.   S.21 LDA 1991 provides a drainage board with powers to serve notice on a person who is liable to do any work in relation to any watercourse, bridge or drainage work and who has failed to do that work. The notice may require the person to carry out such work, providing the obligation to do it would normally fall on that person by reason of 'tenure, custom, prescription or otherwise'.

33.   The obligation must have been in existence before the commencement of the LDA 1991, and it cannot be an obligation 'under an enactment re-enacted in the LDA 1991 or the WRA 1991' (S.21(l) LDA 1991). It is reported that because of these requirements, this section is difficult to invoke successfully.

34.   Under SS.21(4) and (5) LDA 1991, a drainage board may act in default after 7 days' notice, and recover all expenses from the person liable.

35.   In an internal drainage district, the 'drainage board' concerned is the IDB, and elsewhere the Agency (S.21(6) LDA 1991, and see S.107(2) WRA 1991). Under S.8 LDA 1991, the IDB's powers are exercisable concurrently with those of the Agency.

Incidental powers

36.   S.167 WRA 1991 and S.15 LDA 1991 provide powers to appropriate and dispose of spoil from dredging work without making payment for it, and to deposit it on adjacent land. S.155 WRA 1991 gives certain

powers to acquire land, but these powers are exceptional and relate only to accretions of land resulting from land drainage works. The general powers for IDBs to acquire land are to be found in S.62 LDA 1991, and the powers for the Agency in S.154 WRA 1991 and S.37(1)(b) EA 1995.

Directions to Internal Drainage Boards

37.　　In line with the tiered system of land drainage administration described in 1.17 above, the Agency is empowered by S.7 LDA 1991 to give general or special directions to IDBs for the purpose of securing the efficient working and maintenance of existing drainage works and the construction of such new works as may be necessary. This section also requires the Agency's consent if an IDB wishes to undertake work which affects another IDB.

38.　　S.9 LDA 1991 empowers the Agency also to act on behalf of, or in default of, IDBs. Chapter 4 below enlarges upon this.

Power to carry out schemes for small areas

39.　　By virtue of S.18 LDA 1991, the Agency or a county council or unitary authority may carry out schemes for the improvement of small areas of land, where the setting up of a new IDB for such works would not be practicable. The powers are subject to limitations on maximum costs, which must be apportioned between, and are recoverable from, the 'several owners of the lands', subject to a fixed rate per hectare.

Power to make byelaws

40.　　S.66 LDA 1991 provides that IDBs and local authorities other than English county councils may make such byelaws as they consider necessary for 'securing the efficient working of the drainage system in their district or area'. Byelaws for the same purpose and for 'securing the proper defence of any land against sea or tidal water' may also be made by the Agency (S.210 WRA 1991 and Schedule 25, paragraph 5). Local authorities may make byelaws only in relation to ordinary watercourses, and only for preventing flooding or remedying or mitigating any damage caused by flooding.

41.　　In general, byelaws made by the Agency concern main rivers only, and deal with matters such as the erection of fences, disposal of rubbish, excavation affecting the beds or banks of main rivers, erection of jetties and walls, tree planting, use of vehicles on river banks, damage caused by fishing, grazing, etc., destruction of vermin, mooring of vessels, consenting arrangements and similar matters.

Flood warning systems

42.   S.166 WRA 1991 empowers the Agency to provide and operate flood warning systems (see the statutory definitions at the front of this book) and to provide, install and maintain apparatus for this purpose. S.69 of the WA 2003 relates to grants for flood warning schemes.

Tidal flooding and coastal erosion

43.   The Agency, under S.165 WRA 1991, has powers to carry out works for the purpose of protecting low-lying coastal or estuarine land against tidal flooding, both above and below the low-water mark. These powers do not extend to coastal protection against erosion. This is covered by the CPA 1949, under which the coastal local authorities have responsibilities (see Chapter 16 below).

Financing of land drainage functions

44.   The Agency is empowered by S.133 WRA 1991 to raise revenue by requiring payment of its flood defence expenses under the provisions of the Local Government Finance Act 1988. This provides for the issuing of levies apportioned between the relevant local authorities within each local land drainage district. The financing arrangements for land drainage are therefore separate from the Agency's other funds and charges, which come largely from central government grant aid and authorisation fees. S.69 of the WA 2003 relates to grants for flood drainage schemes.

General aims and environmental duties

45.   The principal aim of the Agency is to discharge its functions in a way which 'contributes towards attaining the objective of achieving sustainable development'. The Agency will be guided by the advice of Secretaries of State in this objective, and it must take into account likely costs (S.4 EA 1995). The Agency is also subject to general duties, including general environmental duties, under SS.6–9 EA 1995, as well as specific environmental duties in relation to flood

defence under the LDA 1994. These are discussed in more detail in Chapter 17 below.

46.   In the revised *Management Statement* issued to the Agency in July 2002, ministers stated seven general objectives for the Agency as originally set out in the EA 1995. In addition to objectives related to operating to high professional standards in ways reflecting good environmental practice, these included the following objectives:

(a) to adopt, across all its functions, an integrated approach to environmental protection and enhancement which considers impacts of substances and activities on all environmental media and on natural resources

(b) to work with all relevant sectors of society, including regulated organisations, to develop approaches which deliver environmental requirements and goals without imposing excessive costs (in relation to benefits gained) on regulated organisations or society as a whole

(c) to adopt clear and effective procedures for serving its customers, including by developing single points of contact through which regulated organisations can deal with the Agency

(d) to develop a close and responsive relationship with the public, local authorities and other representatives of local communities, and regulated organisations.

4

Internal Drainage Boards

Functions and constitution

1.   IDBs are independent bodies established in particularly low-lying areas of England and Wales where flood protection and land drainage are necessary to sustain both agricultural and developed land use. There are around 170 IDBs, covering an area of some 1.2 million hectares. A high proportion of this area requires pumped drainage to evacuate water.

2.   The functions of the IDBs and the Agency do not overlap and are specified clearly in the LDA 1991 (Part II) and the WRA 1991, respectively.

3.   Schedule 1 to the LDA 1991 sets out the arrangements for determining the membership of IDBs. The Boards consist of members elected by the agricultural ratepayers for a three-year term. Those so elected must be owners or occupiers of land within the district. The members are elected by occupiers of land in the district by means of a system under which up to 10 votes per person may be cast according to the value of the agricultural land and buildings occupied. This plural voting system gives more votes to those occupiers who contribute higher drainage rates.

4.   Local authorities which contribute towards the costs of the Board are able to nominate members to the Board in numbers commensurate with the proportion of the Board's total income which they contribute. The maximum number of such nominated members provides a bare majority of one over the elected members.

5.   Schedule 2 to the LDA 1991 lays down rules covering the proceedings of the Boards.

6.   SS.2–9 LDA 1991 cover the supervisory role of the Agency with respect to IDBs, and include, among other matters, powers to review the boundary of an internal drainage district and to take over, where necessary, the duties of the Board, both with the approval of the relevant Secretary of State.

The Land Drainage Act 1991

7.　For the purpose of the LDA 1991, IDBs and local authorities are statutorily defined as 'drainage bodies'. PPS25 draws attention to the need to ensure that where flood risk considerations arise, they should always be taken into account by local planning authorities in the preparation of development plans and in the determination of planning applications (see 3.8 above and, more generally, Chapter 10 below). S.105 WRA 1991 requires surveys relating to flood defence functions to be conducted by the Agency in certain areas (see 3.13–3.15 above). These surveys are of particular value in relation to planning matters, but will not dispense with the need for detailed consultation in individual cases.

8.　IDBs have a duty, under S.14 LDA 1991, to exercise general supervision over all matters relating to the drainage of land within their districts. They are also empowered to carry out work on all watercourses within their area except for main rivers, which are the responsibility of the Agency. Although these powers are permissive, in practice most Boards designate certain watercourses in their area upon which to carry out regular maintenance. There is no requirement for a statutory map.

9.　Under Part IVA LDA 1991, inserted by LDA 1994, which repealed SS.12 and 13 LDA 1991, IDBs now have enhanced environmental and recreational duties. These are discussed more fully in Chapter 17 below.

10.　Smaller watercourses will generally remain the primary responsibility of riparian owners (see Chapter 6 below).

11.　Within an internal drainage district, the local authority's permissive powers overlap with those of the IDB. Local authorities' powers are restricted to providing flood protection, and they would not normally become involved in providing a drainage service to agricultural land.

12.　In recent years, substantial development has taken place in many IDB districts, and this has both increased the pressure on the system and often necessitated the maintenance of higher standards. With many IDBs, a significant thrust of their work is now to provide adequate standards of flood protection through pumping stations and channel systems to urbanised areas, including major industrial complexes as seen in areas such as the south bank of the Humber Estuary.

13.　In addition to carrying out construction, maintenance and improvement works in its area, an IDB may:

(a) require or execute default remedial works to maintain proper flows in watercourses (S.25 LDA 1991)

(b) make general byelaws for the efficient working of the drainage system (S.66 LDA 1991)

(c) acquire land, or dispose of it, for the performance of its functions (SS.62 and 63 LDA 1991).

The Boards are empowered under S.23 LDA 1991 to levy a charge of £50 on applicants seeking consents for such works, although this sum may be varied by ministerial order.

14.   IDBs are locally based organisations, and, as such, there is a close relationship between their members, local authorities and ratepayers. The size of their districts is determined by that area which receives benefits from flood protection and drainage works. This benefit can be provided by the IDB exclusively, or in part by the Agency. In respect of the latter, there are financial arrangements between IDBs and the Agency.

The Water Resources Act 1991

15.   Subject to the provisions of the EA 1995, the WRA 1991 covers the functions of the Agency, including flood defence.

16.   Within the WRA 1991, provision is made under S.139 for the IDBs to pay a precept to the Agency. The WA 2003 reinstated the power of the IDBs to borrow to fund contributions to the Agency.

Financial arrangements for Internal Drainage Boards

17.   The income of an IDB is derived from various sources, which may be listed as follows (see Part IV LDA 1991):

(a) Drainage rates are levied on agricultural land and buildings within the drainage district.

(b) Special levies are charged to local authorities (district councils or unitary authorities) which reflect the benefit received by non-agricultural property within the Board's district, including domestic, commercial and industrial buildings and other land such as recreation areas on which agricultural rates would not be levied.

(c) Where watercourses and pumping stations maintained by an IDB transfer and evacuate water which flows through the district from upland areas, then contributions towards the Board's costs can be made by the Agency. This is covered in S.57 LDA 1991.

(d) When developments are proposed within the district, the Board considers what impact, if any, this will have on the drainage system; and if works, either immediate or in the future, are necessary to deal with extra water, then an appropriate contribution can be sought from the developer.

(e) A charge of £50 may be payable by those applying for consent to carry out works under S.23 LDA 1991. This sum may be varied by ministerial order.

(f) With respect to capital improvement schemes, Defra provides grant aid of 25% of the cost where such schemes meet the required benefit criteria (see Chapter 14 below).

18. Under S.139 WRA 1991, the Agency imposes a precept on IDBs with respect to the benefit which the drainage district and its ratepayers received from the Agency's flood defence works.

19. The annual report and accounts of the IDB must be sent to the Secretary of State. At the same time, copies must also be sent to the Agency, and the appropriate county council or unitary authority (Schedule 2, paragraphs 4–5, LDA 1991).

Draft Flood and Water Management Bill

20. This Bill will clarify responsibilities managing flood risk and coastal erosion at a local and national level. The aim is to retain the existing roles and responsibilities where possible, and promote local partnerships, whilst requiring all bodies to cooperate and share information. IDBs will have updated powers to manage the risk of flood and coastal erosion in line with local and national strategies.

IDB Review

21. The Pitt Review expressed concern over the complex arrangements relating to drainage systems and the numerous bodies involved. Defra has led an IDB Review, proposing sub-catchment units that will each be ultimately led by an IDB, and that these new sub-catchment boards should be in place by April 2013. This recommendation follows an independent review of IDBs in 2005.

5

Local authorities

Outline of powers

1. Local authorities with land drainage powers, as defined in the LDA 1991, are county councils, county borough councils (in Wales), borough councils (in England), metropolitan and non-metropolitan district councils, London boroughs, and the Common Council of the City of London. These authorities are included in the general term 'drainage bodies' defined in S.72 LDA 1991, together with the Agency, IDBs and any other body having the power to make or maintain works for the 'drainage of land'.

2. Following local government reorganisation in both England and Wales, a greater number of unitary authorities now exist. Broadly, though other arrangements may be provided for, these combine the powers of county and district councils. In Wales, all authorities are now unitary authorities, but may be designated either as county councils or county borough councils. In England, some unitary authorities are being created in addition to the metropolitan districts and London boroughs, which were already unitary authorities; elsewhere, the two-tier structure of county councils and district councils (the latter sometimes called borough councils) remains (see Local Government Act 1992 and Local Government (Wales) Act 1994).

3. The powers of local authorities are given principally in SS.14–18, 20(2), 25, 26 and 60 LDA 1991. These generally relate to flood prevention, maintaining flows in watercourses and the making of byelaws. Some additional powers are covered in Part V LDA 1991 (Miscellaneous and Supplemental Provisions). Local authorities also have extensive and important powers as planning authorities. These are largely dealt with in Chapter 10 below, though specific comment is made on planning obligations in 5.30 below.

4. The following paragraphs follow the basic distinction between district and county council functions, while recognising the existence of unitary

28

authorities combining these powers. However, the London unitary authorities are treated separately, as in some, though not all, cases, separate provision is made for them in legislation.

English district councils and unitary authorities and Welsh unitary authorities

Principal powers

5. As far as the prevention, mitigation and remedying of flood damage is concerned, S.14 LDA 1991 confers powers on English district councils and unitary authorities of the same kind as those given to the Agency and IDBs.

6. Before work is carried out under the powers conferred by S.14 LDA 1991, consent is required from the Agency, as provided for in S.17 LDA 1991.

7. As described in 3.20–3.25 above, the Agency is provided with control powers by means of a mandatory 'prior consent' procedure under S.23 LDA 1991 for ordinary watercourses, and under S.109 WRA 1991 for main rivers. These powers give control over the erection or alteration of any structure or culvert which is liable to obstruct the flow of such watercourses.

8. Where obstructions do occur in ordinary watercourses, local authorities have powers to secure a proper flow under S.25 LDA 1991, although they must inform the Agency of any action to be taken under this section (S.26 LDA 1991). Obstructions may also be dealt with by local authorities under public health powers, or by byelaws made under S.66 LDA 1991.

The Land Drainage Act 1991

9. Powers exist to maintain and improve existing ordinary watercourses and to construct new works (S.14(2) LDA 1991 and see 3.17 above).

10. A local authority may, if it wishes, carry out drainage works, other than on main rivers, 'for the benefit of its area, in lands outside that area' (S.14(3) LDA 1991).

11. A local authority may carry out the normal land drainage practice of spreading spoil on adjacent land while carrying out watercourse improvement works (S.15 LDA 1991).

12. Any county council or unitary authority may undertake drainage work for any person at that person's expense (S.20 LDA 1991).

13.   Any local authority may serve notice on persons requiring them to carry out necessary works to maintain the flow of ordinary watercourses (S.25 LDA 1991).

14.   The consent of the Agency is required before the exercise of any of the powers listed above (SS.17 and 26 LDA 1991).

15.   By virtue of S.59 LDA 1991, grants may be obtainable from Defra as set out in the Ministry's memorandum relating to grants (see Chapter 14 below).

16.   A local authority may make byelaws to secure the efficient working of the drainage system in its area (S.66 LDA 1991). Schedule 5 to the Act gives further guidance on this aspect.

Public Health Act 1936

17.   Some of the powers given to district councils and unitary authorities in the PHA 1936, and Part III of the Environmental Protection Act 1990 concerning statutory nuisances, have relevance to land drainage matters. These are summarised in the following paragraphs. There was an amendment to the Public Health Act in 1961, but this does not relate to watercourses, ditches or ponds.

18.   Polluted pools, ponds, ditches, gutters or watercourses which cause a nuisance or give rise to conditions prejudicial to health will be statutory nuisances. A statutory nuisance will also arise where non-navigated watercourses become so choked or silted up as to obstruct or impede the proper flow of water and cause a nuisance or be prejudicial to health, but only the person by whose act or default the latter arises can be liable (S.259 PHA 1936). Doing nothing to prevent a natural stream from being obstructed by natural causes, though, will not amount to an 'act or default' (*Neath RDC* v. *Williams* [1950] 2 All ER 625), though there is some uncertainty whether the decision in *Leakey* v. *National Trust* (1980) QB 485 affects this principle (see 6.5 below).

19.   Similar powers for district councils and unitary authorities to deal with problems resulting from drainage, filth, stagnant water or matter which is likely to be prejudicial to health also exist. Powers are conferred to execute works, including maintenance or improvement works (S.260 PHA 1936).

20.   A district or unitary authority may also take action against a neighbouring authority to remedy a nuisance caused by a boundary watercourse, or a watercourse near a boundary with an adjoining district (S.261 PHA 1936).

21.   District or unitary authorities may require a developer to fill, partially fill, cover over, pipe in or culvert any ditch or watercourse (other than the main river) that runs through or abuts land to be developed. In dispute, either party may appeal to a court of summary jurisdiction (S.262 PHA 1936).

22.   It is unlawful to culvert or cover any watercourse except in accordance with plans and sections to be submitted to and approved by the district or unitary authority (S.263 PHA 1936). However, the authority cannot require the culvert to be oversized, to deal with future flows, as a condition of approval. It can, though, request such a provision and reimburse any additional costs reasonably incurred.

23.   District or unitary authorities may require landowners to repair, maintain and cleanse culverts in watercourses in, on or under their land (S.264 PHA 1936).

24.   District or unitary authorities may also contribute the whole or a part of the cost of works in relation to watercourses and other work covered by the above sections of the PHA 1936 (S.265 PHA 1936).

25.   District or unitary authorities may not exercise the above powers within the jurisdiction of a drainage body until that body has been consulted, unless they have instituted proceedings for statutory nuisance (S.266 PHA 1936).

26.   Although all local authorities are subject to environmental duties under S.61B LDA 1991, these only relate to the exercise of functions under the LDA 1991 and not more generally to the exercise of any powers affecting land drainage. Thus, it may be possible for a local authority not to be bound by specific environmental duties where it exercises powers under enactments such as the PHA 1936 noted above. This observation applies equally to county councils and London authorities.

Highways Act 1980

27.   A metropolitan district council or unitary authority, acting in its capacity as a highway authority by virtue of the Local Government Act 1985, also has the powers available to a county council referred to subsequently in 4.42–4.51 below.

28.   In connection with culverts, SS.100 and 101 HA 1980 are also relevant, providing powers for highway authorities to drain water from highways, or to fill in or pipe roadside ditches. S.23 LDA 1991 and S.109 WRA 1991 should also be noted, as respectively they require the consent in writing of IDBs or the Agency in all cases where the culverting of any watercourse is proposed (see 3.21–3.25 above).

Local Government Act 1972

29.   Section 138 of the Local Government Act 1972 enables any council to incur expenditure to avert, alleviate or eradicate the effects or potential effects of any emergency or disaster. However, these powers cannot be used to carry out works on a main river, or used as alternative powers to those available under S.14 LDA 1991.

Section 106 planning obligations

30.   A situation that will require co-operation between the local authorities and the Agency occurs where a major development increases the run-off into a watercourse prone to flooding. Here, an obligation undertaken under S.106 of the Town and Country Planning Act 1990 (TCPA 1990) (see 10.27–10.28 below) may be sought, in which the developer undertakes to pay for the necessary drainage improvements before planning permission is granted. The initiative would generally be taken by the district council, since the county council's planning powers relate to structure plans and strategic matters, apart from minerals, waste disposal and its own developments. The consent of the downstream landowner to the carrying out of such works is an essential prerequisite to such an agreement.

English county councils and unitary authorities

Functions

31.   A county council in England has two distinct interests in land drainage matters, i.e. as a drainage body and as a highway authority. Powers given to 'county councils' as referred to in the following sections largely extend also to unitary authorities, where these exist. In June 2008, the Government announced that local authorities will take responsibility for surface water management, including Surface Water Management Plans. This followed the Agency's new strategic overview role in England for all sources of flood risk.

Powers as a drainage body

32.   The LDA 1991 confers powers on county councils which in some respects are similar to some of the powers given to district councils. Important powers conferred by S.14 LDA 1991 on district councils, however, are not available to the remaining county councils in England,

except at the request of the district council or where the district council fails to exercise the powers. SS.16, 18 and 20 LDA 1991 confer powers on English county councils which are not available to district councils (see 5.37 and 5.39 below).

33. The permissive powers concerned with land drainage are given in SS.14, 18, 20, 25, 60, 62 and 64 LDA 1991.

34. If the powers contained in SS.14 and 15 LDA 1991 are not exercised by the relevant district council, the powers may be exercised by county councils by virtue of the provisions of 16(1) LDA 1991 either:

(a) at the request of the district council or
(b) after not less than 6 weeks' notice by the county council to the district council.

35. Although counties have in many instances assisted district councils in land drainage schemes, they have rarely been willing to act in default of a district council. Assistance may be in the form of preparation of, or advice on, such schemes.

36. Grant aid may be available to county councils or unitary authorities from Defra for schemes carried out under SS.18 and 20 LDA 1991 in accordance with S.59 LDA 1991. The principal role of the county council in land drainage has therefore been to obtain co-operation between land owners or occupiers who would benefit from a scheme, and to obtain a Defra grant (see Chapter 14 below).

37. S.20 LDA 1991 enables a county council or unitary authority to execute such land drainage schemes at the request of owners or occupiers who would benefit from the scheme.

38. Powers under S.25 LDA 1991, as conferred on IDBs (see details in 3.29–3.31 above) are also available to a county council, thereby enabling the council to serve notice on the categories of persons listed below, requiring them to carry out necessary works to maintain the flow. The categories are:

(a) any person having control of the part of the watercourse where any impediment occurs
(b) any person owning or occupying land adjoining that part
(c) any person to whose act or default the condition is due.

Details of the procedure to be followed in serving notice are also included in S.25 LDA 1991.

39. S.18 LDA 1991 enables county councils to make a scheme for drainage of small areas (see 3.39 above) and to apportion the expenses among the owners of the areas (usually agricultural land) to which the

scheme relates. Any proposal to exercise these powers requires prior consultation with the Agency.

40.   Provision is made for the cost of schemes for the drainage of small areas to be apportioned up to a maximum of £50 per hectare. This maximum figure, coupled with administrative complexity, tends to limit the utility of S.18 LDA 1991. Provision is available for the Secretary of State to exempt a scheme from this limit if 'it is urgently required in the public interest'. Otherwise, the limit may be varied by ministerial order.

41.   Engineers have reported that, in practical terms, the willingness of the relevant Secretary of State to agree to a revised figure, where justifiable, has enabled some schemes to be carried out.

Drainage powers as a highway authority

42.   As a highway authority for all roads within its area, except trunk roads, there is a responsibility on a county council or appropriate unitary authority to keep the roads free from flooding and to make provision for run-off from highways in a proper manner, using powers now contained in the HA 1980. The Minister of Transport has the same powers in relation to trunk roads. The following provisions of the Act are applicable.

43.   S.100 HA 1980 gives the power to prevent water flowing on to a highway; the power to drain water from a highway; the requirement to compensate a person who suffers damage because of the drainage works; and the power for a highway authority to exercise certain powers and duties which are also exercisable by sewerage undertakers under the WIA 1991.

44.   The highway authority may also fill in or culvert roadside ditches where they are a danger to highway users. This is subject to a requirement to obtain written consent from the drainage board under S.23 LDA 1991 where culverting is proposed (S.101 HA 1980).

45.   The Secretary of State may authorise the highway authority to divert a navigable watercourse to facilitate highway works (S.108 HA 1980).

46.   A non-navigable watercourse may, after consultation with every local council in the area, be diverted. Certain other works on all watercourses may also be undertaken (S.110 HA 1980).

47.   Highway authorities may also require adjoining occupiers to prevent water from roofs from falling onto persons using a highway or from flowing over footways (S.163 HA 1980).

48.   Road drains are vested in county councils or unitary authorities, and the right to discharge highway run-off into sewers, if established before 1 April 1974, is continued (S.264 HA 1980).

34

49.   Surface water drains may be discharged into inland or tidal waters, but with an obligation to pay compensation to an owner or occupier of land who suffers damage (S.299 HA 1980).

50.   Highway authorities must obtain consent from the Agency or other drainage body before any watercourse is used or interfered with, or before works are carried out on any watercourse or drainage works (S.339 HA 1980).

51.   Under S.115 WIA 1991, a highway authority can by agreement with a sewerage undertaker arrange to discharge water from the highway to the undertaker's sewers. An undertaker may similarly use any highway drain to drain surface water from premises or streets. Neither party can unreasonably refuse to enter into an agreement or insist on unreasonable terms; appeal is to the Secretary of State. S.146 WIA 1991 prevents an undertaker from requiring any payment by a highway authority. This would probably not prevent an agreement being made to share costs where it was of mutual benefit to do so.

52.   Whether a drain is properly described in law as a highway drain or a land drain, and thus whether or not it is the responsibility of the county council as the highway authority, is considered in 12.25–12.27 below.

Planning matters

53.   A county council may wish to become a party to a planning obligation in order to safeguard its interests as a highway authority, e.g. to ensure that works are carried out to reduce the risk of flooded highways or to improve the capacity of culverts under the highway (see also 5.30 above).

The London authorities

54.   Most drainage powers given to local authorities are also possessed by the London authorities – the London borough councils. However, some powers are specific, and the position with relation to the London authorities is outlined here for convenience.

55.   For the purpose of land drainage, the LDA 1991 confers powers on London boroughs and the Common Council of the City of London under the definition of local authorities. However, there are sections in the Act where specific mention is made of the London boroughs as there is no county council involvement within Greater London. The permissive powers over ordinary watercourses may be exercised by London boroughs with the prior consent of the Agency in accordance

with SS.17 and 26 LDA 1991. Powers specifically conferred on the London boroughs are given in SS.10, 15, 16, 20, 55, 57, 61 and 62 and Schedule 2 LDA 1991. These powers are summarised in the following paragraphs. With the exception of SS.10, 15 and 57, these provisions apply equally to the Common Council of the City of London.

56.   By virtue of S.10 LDA 1991, the council of a London borough may make an application to the Agency to exercise within its area the default powers conferred by S.9 LDA 1991 on the Agency. If the Agency refuses, then the council has the right to appeal to the Secretary of State.

57.   An IDB or local authority may enter into an agreement to pay the council of a London borough for the disposal of spoil removed from a watercourse while drainage works for flood prevention or the mitigation of any damage caused by flooding are carried out (S.15 LDA 1991).

58.   If powers conferred on a London borough by virtue of S.14(1) LDA1991 are not exercised by that council, they may be exercised by the Agency after it has given at least 6 weeks' notice in writing to the council (S.16(2) LDA 1991). Any expenses incurred by the Agency are recoverable from that council by the Agency summarily as a civil debt. If the council appeals against the notice, the Agency may not exercise the powers under S.16 LDA 1991 until confirmation is given by the Secretary of State.

59.   By virtue of S.20 LDA 1991, a London borough may enter into an agreement with any person to carry out drainage works on watercourses within its area. Such works will be at that person's expense and must be works which that person is entitled to carry out and maintain.

60.   A council of a London borough and the Common Council of the City of London may borrow for the purposes of this Act (S.55 LDA 1991).

61.   A council of a London borough may appeal to the Secretary of State against any excessive contributions by the Agency to the expenses of an IDB (S.57 LDA 1991).

62.   By virtue of S.61 LDA 1991, the expenses of a council of a London borough under the LDA 1991 are to be defrayed as general expenses or special expenses charged on such parts of the borough as the council thinks fit.

63.   S.62 LDA 1991 deals with the powers of IDBs and local authorities to acquire land. A London borough may be authorised by the Secretary of State to purchase land compulsorily to exercise the powers given under SS.14–17 LDA 1991 and byelaws of the council made under S.66 LDA 1991 for preventing flooding or for remedying or mitigating any damage caused by flooding.

6

Riparian owners

Riparian rights and duties

1.	The proprietor of land on the banks or under the bed of a natural watercourse is entitled to the enjoyment of what are commonly known as 'riparian rights', based on common law. Where a channel is not of natural origin, the same rights may not apply; further comment is made in 6.31–6.33 below. For these rights to arise, it is necessary for the water to flow in a defined channel, which may be over or below ground. The flow may only be intermittent for the channel to be a watercourse at common law (*Stollmeyer* v. *Trinidad Lake Petroleum Co.* (1918) AC 485). Percolating water and water which lies or flows generally over the surface is not the subject of riparian rights.

2.	Lord Wensleydale, in *Chasemore* v. *Richards* (1859) 7 HL Cas 349, said, 'It has been settled that the right to the enjoyment of a natural stream of water on the surface *ex jure naturae* belongs to the proprietor of the adjoining land as a natural incident to the right to the soil itself; and that he is entitled to the benefit of it, as he is to all the other advantages belonging to the land of which he is the owner. He has the right to have it come to him in its natural state, in flow, quantity and quality, and to go from him without obstruction, on the same principle that he is entitled to the support of his neighbour's soil for his own in its natural state. His right in no way depends on prescription or the presumed grant of his neighbour, nor from the presumed acquiescence of the properties above and below.'

3.	The engineer will be required to deal with various types of riparian owner, including the agricultural landowner, the industrial or commercial landowner, the owner of dwelling premises and the developer, i.e. an owner of land on which it is intended to carry out development such as housing, industrial units, etc.

4.	The Agency website lists responsibilities for riparian owners and has published a booklet entitled *Living on the Edge*. The owner must pass flows forward without obstruction, pollution or diversion. The owner must maintain the bed and banks of the watercourse, keeping it free of any debris, including debris that may be washed into the watercourse, including any structures such as culverts, screens and weirs. The WRA 1991 and associated byelaws require owners to apply to the Agency for formal consent to works in, over, under or adjacent to main rivers, referred to as 'flood defence consents'. Under the LDA 1991, owners also need consent from the Agency to construct a culvert or flow control structure (e.g. a weir) on an ordinary watercourse.

5.   In *Leakey* v. *National Trust* (1980) 1 All ER 17, a landowner was held liable for naturally occurring slides of soil, rocks, tree-roots and other debris from a bank onto a neighbouring property. In deciding the case, the court stated that there existed a general duty on occupiers in relation to natural hazards occurring on their land, so that where the hazard encroached or threatened to encroach on another's land, there was a duty to do all that is reasonable in the circumstances to prevent or minimise the risk of foreseeable damage to the property of others. It is notable that the case was not concerned specifically with drainage matters, and is in direct conflict with previous authorities, but it may be indicative of a change in thinking about the proper extent of drainage responsibilities. Hence, a riparian owner or occupier may now be liable for any nuisance caused if he fails to remedy any defect in a bank or sea wall within a reasonable time after he became, or ought to have become, aware of it (J.H. Bates, *Water and Drainage Law*, para. 2.39). The uncertainty generated by the *Leakey* case has not been resolved by subsequent court decisions (though see 6.12 below), and the following discussion states the traditional view of the rights and duties of riparian owners while recognising that the law may have been implicitly changed as a result of the decision.

6.   A riparian owner must accept alterations to the state of a natural stream, which changes could not be practicably avoided as a result of work carried out under statute (*Provender Millers (Winchester) Ltd* v. *Southampton County Council* (1940) 1 Ch. 131). The extent to which a planning permission similarly provides a defence against an action in civil law is not presently settled in law, but it would appear that, unless the permission changes the character of a locality, successful civil claims will remain possible (*Wheeler* v. *Saunders* (1995) 2 All ER 697; *Hunter and others* v. *Canary Wharf Ltd* (1995) *The Times*, 13 October). In general, however, the issue will be determined according to the general rules of interpreting the statute in question.

7.   Traditionally, a riparian owner is not liable for damage, e.g. by erosion, caused to adjoining land by virtue of the natural action of water on the land adjoining or downstream, provided that there is no negligence or wilfulness involved (*Rouse* v. *Gravelworks Ltd* (1940) 1 KB 489).

8.   The proprietor of land has a natural right to collect and discharge the surface water run-off from his land onto adjoining lower land, even where he has created an increase in flow by an improved system of drainage or by other means (*Durrant* v. *Branksome Urban District Council* (1897) 76 LT 739). This right does not permit an owner or an authority to act in a wilful manner, as where a highway authority deliberately drained its land onto the land of adjoining occupiers (*Thomas* v. *Gower Rural District Council* (1922) 2 KB 76). However, a lower occupier of land is under no obligation to receive water, and may take steps to pen back the flow of water, even if this causes damage to higher land. Moreover, the right to prevent water entering land is not an absolute one, and will only allow the lower landowner to take reasonable steps to prevent the entry of the water (*Home Brewery plc* v. *Davis and Co.* (1987) 1 All ER 637).

9.   It follows that increased natural flows from upstream must be accepted and dealt with by the proprietor of land downstream. The flow may flood the downstream land unless adequate drainage works are carried out by him to prevent it. However, if flooding is caused by inadequate capacity further downstream, the downstream riparian owner has no common law duty to improve the watercourse.

10.   The overall situation might be described as a 'sequential system of rights to discharge'.

11.   While riparian owners therefore have the responsibility to accept the flow, it might be unfair to expect them to carry out extensive and important works to deal with substantial additional flow, e.g. from a proposed development. A developer wishing to obtain planning permission will need to satisfy the planning authority that adequate drainage arrangements will be made.

12.   An owner who collects and keeps on his land anything which is likely to do mischief if it escapes, e.g. a water reservoir, will be liable for any reasonably foreseeable damages caused by the escape (*Rylands* v. *Fletcher* (1868) 19 LT 220; *Cambridge Water Co.* v. *Eastern Counties Leather* (1994) 1 All ER 53). But an owner who stores water on his land, exercising reasonable care, is not liable for an escape of water which injures his neighbour, if the escape is caused by a factor beyond his control, such as a storm, the impact of which is practically, but

not physically, impossible to resist (*Nicholls* v. *Marsland* (1875) 33 LT 265). Another exception to the basic rule in *Rylands* v. *Fletcher* is where the owner has no control over the reservoir or knowledge of the conditions which led to the escape (*Box* v. *Jubb* (1879) 3 Ex. D-76; *Nield* v. *London & North Western Railway Co.* (1874) LR 10 Ex. 4).

13.   If a watercourse overflows on to his land, the riparian owner has no remedy unless he is able to show that the injury is due to a wilful or unlawful act of another riparian owner, either upstream or downstream (*Mason* v. *Shrewsbury and Hereford Railway Co.* (1871) 25 LT 239; see also 6.8 above).

14.   Alternatively, it would be necessary for him to show that the damage has been caused by the failure or neglect of some other body or person with responsibilities to maintain the channel (*Harrison* v. *Great Northern Railway Co.* (1864) 3 H&C 231; *Smith* v. *River Douglas Catchment Board* (1949) 113 JP 388; *Sedleigh-Denfield* v. *O'Callaghan* (1940) 3 All ER 349).

15.   Although public authorities are not normally liable for not exercising lawful duties where the duty is owed to the public as a whole rather than to individuals (e.g. see 13.34 below), there is a general duty upon both land drainage bodies and riparian owners not to take an action which would have the effect of exacerbating damage which would have been suffered if no action was taken (*East Suffolk Catchment Board* v. *Kent* (1940) 4 All ER 527).

16.   From the riparian owner's point of view, problems may arise, owing to the need to reach agreement, where opposite banks of a watercourse are in different ownership, since, in spite of the riparian duty to maintain flow, different owners can adhere to different standards of maintenance. In some cases, this disparity can result in improvement to only one side of the watercourse, with a resulting increased risk of flooding on the other, and the possibility of consequent legal liability (*Menzies* v. *Breadalbane* (1828) 3 Bli. NS 414).

17.   Problems arise less frequently where there is common ownership of both sides of a watercourse. Such problems can be minimised further as a result of the eligibility for grant aid to riparian owners (though not lessees) towards the cost of approved improvement schemes (on grant aid generally, see Chapter 14 below).

18.   A riparian owner is entitled to protect his property from water that seeps through natural or artificial banks and may make repairs where seepage threatens the structural integrity of the defence. This may be by building a defence against flooding, provided that such defence is not built in the channel so as to cause obstruction. Such works may

constitute development as an engineering or other operation for which planning permission is required under TCPA 1990 (see Chapter 10 below). Consent must also be obtained from the IDB (or otherwise from the Agency; see 3.21 above), and the defence works must not cause actual injury to a neighbour's property or rights (*Bickett* v. *Morris* (1866) 30 JP 532).

19.   In *Hanbury* v. *Jenkins* (1901) 2 Ch. 401, it was stated that a riparian owner may place stakes and wattles on the soil of a river to prevent erosion by floods, and to make pens to prevent cattle from straying. However, he must now obtain planning permission for revetment works and obtain consent from the Agency where a main river is involved. In *Hudson* v. *Tabor* (1877) 2 QBD 290, it was held that a riparian owner was under no duty in the absence of some specific obligation to keep his banks in repair.

20.   There is a 'natural' right of support to land from adjoining land, but the obligation is negative, i.e. to refrain from any act which will diminish support. The owner is not obliged to take active steps to maintain the things that give support. For instance, there is no legal remedy against subsidence caused by the natural erosion of a river bank (on the right of support of property by water generally see *Stephens* v. *Anglian Water* (1987) 3 All ER 379).

Nuisance

21.   The riparian owner must not only exercise due care in the above matters but must also not cause or perpetuate a nuisance. In the Court of Appeal in *Pemberton and another* v. *Bright and others* (1960) 1 All ER 792, the first defendants were held liable for flooding caused by a blocked culvert constructed by the highway authority, without a protecting grid, on the grounds that they had continued the nuisance. The Court applied the principle of law laid down in *Sedleigh-Denfield* v. *O'Callaghan* (1940) 3 All ER 349, that an occupier of land on which a nuisance has been created by another person is liable if he allows the nuisance to continue.

Obstruction to flow — consent procedure

22.   Riparian landowners or householders are frequently unaware of the requirements of S.23 LDA 1991 by which the consent of the drainage authority is required before the construction of, or alteration to, any mill dam, weir or similar obstruction or culvert in an ordinary

watercourse. As far as a culvert in particular is concerned, it is necessary for a proposed culvert to be likely to affect the flow before the requirement for consent will apply (see 13.13 below in this respect; see also *Dear* v. *Thames Water* in 13.34 below for an illustration of a riparian owner being unwittingly liable for a culverted watercourse).

23.   Where the proposed works (which may include a culvert or any of the 'obstructions' referred to above) concern a main river across or alongside their land, riparian owners commonly appreciate requirements for consent required by SS.109 and 110 WRA 1991.

Disputes regarding ditches

24.   If a dispute occurs between adjoining landowners over the maintenance of ditches, the matter can be taken to the Agricultural Lands Tribunal (SS.28–31 LDA 1991). Additional information concerning ditches, ownership problems, etc., is given in 8.3–8.6 below.

Disputes between neighbours

25.   Although there are powers for statutory authorities to organise or carry out work where several riparian owners are involved, it is sometimes difficult for a small number of them, or perhaps one alone, formally to request such work. The difficulty arises because there is a reluctance to force issues with unwilling neighbours, in spite of the considerable rewards, particularly for farmers, which can result from the improved drainage of agricultural land. In general, there is a tendency for such difficulties to be solved by agreement once the drainage problem becomes sufficiently serious (see 5.37 above, and 8.9 below).

Powers of entry

26.   Riparian owners generally accept the need for the wide powers of entry which are possessed by land drainage authorities, and usually little difficulty is experienced in reaching agreement without the use of such powers. Requirements for notice are contained in S.64 LDA 1991.

Problems regarding grant aid

27.   Riparian owners sometimes encounter difficulty in obtaining grant aid from Defra for watercourse improvement schemes. Grant aid is

not, however, made available on the same basis as for ditching and tile drainage schemes. An IDB or a county council or unitary authority may carry out work on behalf of a landowner at his expense under S.20 LDA 1991. The landowner's expense can be reduced by any government grant made under S.59(6) LDA 1991. Any local authority may contribute to works by drainage bodies under S.60 LDA 1991.

28.   Many councils are not prepared to exercise the above powers. Where the powers are exercised, it has also been reported that the resulting costs to the riparian owners can be excessive. This view is, however, necessarily somewhat subjective.

29.   Some of these problems, like so many others associated with land drainage, fall largely into the group which might be described as 'problems of willingness to pay'.

Flooding emergencies

30.   In a case arising from flooding, the riparian owner or householder will rely heavily on the assistance of the local authority, which may provide such assistance under Section 138 of the Local Government Act 1972, e.g. for the provision of sand bags, pumps, driers, etc. (see 5.29 above).

Artificial watercourses

31.   An artificial channel made for a particular purpose will not give rise to riparian rights if it is needed only for the temporary circumstances of that use. Riparian rights would not normally attach, for example, to a mill race used only to serve the mill. Rights to discharge to the mill race or to receive or use water from it may, however, have been acquired by grant or prescription even though the mill may have fallen out of use.

32.   A watercourse of a permanent character may have been constructed originally, and subsequently used in all respects as a natural watercourse, when riparian owners will have acquired the same rights as if it had been a natural stream. Generally, if there is no record of the construction of the channel, and it has not been used for some private object, riparian rights will have been acquired. If it is determined that riparian rights have not been acquired, then the question arises as to what easements in the flow of water have been granted either expressly or by use, if at all. For the law relating to easements, reference can be made to *Gale on the Law of Easements* (18th edn, 2008).

33.   A person diverting a stream into a new and artificial channel for his own convenience must make it capable of carrying off all the water

which may reasonably be expected to flow into it, irrespective of the capacity of the old and natural channel. It is the duty of anyone who interferes with the course of a stream to see that the works which he substitutes for the natural channel are adequate to carry off the water brought down even by extraordinary rainfall, and if damage results from the deficiency of the substitute which he has provided for the natural channel, he will be liable (*Greenock Corporation* v. *Caledonian Railway Co.* (1917) AC 556; *R.* v. *Southern Canada Power Co.* (1937) 3 All ER 923).

Obstruction and flooding

34.   A riparian owner is under no common law duty to clear a watercourse which becomes silted or obstructed through natural causes. Under statute law, however, drainage authorities may require and enforce riparian owners to carry out such works under the LDA 1991 and PHA 1936 (see 5.9–5.26 above). A riparian owner is not liable if other land is consequently flooded (but see 13.26 below for natural obstruction of a culvert). An owner cannot remove obstructions which have through time become embedded and form part of the bed if the result is to increase the flow downstream. Action can be taken by drainage authorities under the LDA 1991 or S.259 PHA 1936.

35.   A riparian owner may construct flood banks on his own land to protect it from flooding if he does not thereby cause or increase the severity of flooding to other property (*Menzies* v. *Breadalbane* (1828) 3 Bli. NS 414). In the case of an extraordinary flood, a riparian owner may turn floodwater from his land without regard to the consequences if he is acting to ward off a common danger and not merely protecting his own property from floodwater (*Nield* v. *London & North Western Railway Co.* (1874) 44 LJ Ex. 15). However, consent may be necessary from the Agency or the local authorities under the provisions of the WRA 1991, the LDA 1991 or the Land Drainage Byelaws.

7

Differentiation between land drainage and sewerage

Difference between a public sewer and a watercourse

1. The important differences between public sewers and watercourses are not always appreciated, particularly by those who advocate that these two very different kinds of water channels should be treated in a like manner under the law. It is therefore necessary to distinguish between 'public sewers' as presently defined for the purpose of the WIA 1991 and 'watercourses' as defined for the purposes of the WRA 1991 and LDA 1991.

2. A public sewer will be vested in a sewerage undertaker who is liable to maintain, cleanse and empty it. The function of such a sewer must include the drainage of buildings, and the public will have a right to discharge thereto (subject to the provisions of the WIA 1991).

3. A watercourse, on the other hand, is significantly different in law in that it is seldom vested in a sewerage undertaker. It has a natural source, and, in addition to statutory provisions, it is subject to common law riparian and prescriptive rights, including the right to a flow of water in its natural state, unaltered in quality and undiminished in quantity. Abstractions from watercourses and discharges of effluent to them are governed by the WRA 1991.

4. A 'watercourse' is defined in S.72(l) LDA 1991 as including 'all rivers and streams and all ditches, drains, cuts, culverts, dykes, sluices, sewers (other than public sewers within the meaning of the WIA 1991) and passages, through which water flows'. It has been held that, in relation to the similar definition of watercourse contained in the WRA 1991, the words 'through which water flows' applies only to sewers and passages, and not to the other terms used. Thus, for example, a dry river bed will remain in law a 'watercourse' even where it has completely dried up, but a sewer will only be a watercourse where water flows through it (*R. v. Dovermoss Ltd* (1995) Env. LR 258).

5.   The distinction between a 'watercourse' and a 'public sewer' is therefore not simply one of nomenclature but materially affects the rights of individuals and the requirements to comply with statutory controls. Before 1876, a 'watercourse' could change its status to become a 'sewer' consequent upon the discharge of sewage into it, but such a change is no longer permissible in law. This circumstance arises because, since the Rivers Pollution Prevention Act 1876, it has been an offence to discharge untreated sewage into a stream (see S.85 WRA 1991 for the current law on this). It is not open to a sewerage undertaker to exercise its powers relating to public sewers with respect to work on watercourses.

6.   A watercourse, therefore, is more than a physical entity. It is also a complex and abstract collection of legal rights and duties, relating to landowners and to the land itself. For example, there is a natural right for the owner of a piece of land to have water, which falls on (or arrives at) his own land, discharged to the contiguous lower land of his neighbour, even where the owner of the higher piece of land causes an increase in flow by his works, provided always that such works are lawful (*Gibbons* v. *Lanfestey* (1915) 84 LJPC 158).

7.   In the case of *Taylor* v. *St. Helen's Corporation* (1877) 6 Ch. D. 264, a watercourse was defined as meaning either:

(a) an easement or right to the running of water (involving various parties, including the riparian owner)
(b) the channel itself or
(c) the land over which the water flows.

This definition is that used for the purposes of civil law controls, and should not be confused with the statutory definitions for a watercourse in the WRA 1991 and LDA 1991 as clarified in the *Dovermoss* case (see 7.4 above).

8.   A problem which appears to have been more prevalent in the period which immediately followed the implementation of the Water Act 1973 concerns some watercourses which pass through built-up areas. Some local authorities have claimed that certain of these channels must be public sewers, simply because the drainage scheme under which they were constructed was carried out using public health powers, rather than powers under land drainage legislation. While this begs the question as to whether or not the appropriate powers were used for the scheme, there is no statutory or common law precedent for such a claim, as the legal definition of a 'public sewer' has been specified in particular statutes and by further interpretation in common law (see 7.18–7.27 below for further details).

9.   It is also claimed that in some cases the responsibility for certain old sewer ditches and dykes has not been passed to the sewerage undertaker although they are public sewers. Whether or not they are public or private sewers is legally a matter of fact. There are well-established legal criteria for identifying public sewers, described below, and the criteria should be carefully applied in each case.

10.   It is desirable that legal advice is sought in doubtful cases, as the circumstances in a particular case will rarely be exactly comparable with those which gave rise to an earlier legal precedent in common law. As an example, in the case of *British Railways Board* v. *Tonbridge and Malling DC* (1981) 79 LGR 565, it was ruled that an existing culvert under a railway line at Tonbridge was not a public sewer despite the fact that the volume of water it carried had been increased by large-scale housing development. Hence, in this case, British Rail failed in its appeal court bid to make Tonbridge and Mailing District Council responsible for the culvert (but see also *Hutton* v. *Esher Urban District Council* (1974) Ch. 167, where it was held that a channel might be constructed either as a sewer for the drainage of buildings or as a watercourse to carry the natural discharge from the area).

11.   A similar ruling to that described in *British Railways Board* v. *Tonbridge and Malling District Council* would not necessarily have been applicable if the proposal was for a new culvert to be constructed under the railway by a developer, highway or drainage body, as part of a new development.

Guidelines for differentiation

12.   In cases of doubt, a 'watercourse' which is suspected in reality of being a 'public sewer' should not be so regarded unless it can be identified by means of investigation on the lines indicated below. If the watercourse is shown as a public sewer on the statutory map of such sewers, which sewerage undertakers are required to keep under S.199 WIA 1991, this would certainly be persuasive as to its character, but not necessarily conclusive.

13.   If the conduit in question is not already shown on the public sewer map, the burden of proof should be regarded as resting upon the person who asserts that it is a public sewer. There has been ample time since 1876 (since when it has not been possible to change the status of a watercourse to that of sewer; see 7.5 above) for action to have been taken to enter the conduit on the sewer map.

14.   Similarly, there has been ample time since the date of operation of the PHA 1936 (since when new discharges of surface water from buildings and yards have not been sufficient to change the status of a watercourse to a public sewer) for the conduit to be recorded on the map.

15.   Nevertheless, there will be cases where errors or omissions have occurred, where the rules for identification of a public sewer will need to be applied.

16.   While the identification of a public sewer can sometimes be somewhat complex, it is usually possible to arrive at an answer, in most cases, fairly readily. The following paragraphs are not intended to present a fully exhaustive review of all the legal aspects, but the methods of identification offered are reasonably comprehensive and sufficient for most practical purposes. They include all the factors with which engineers will be concerned where a reasonably rapid assessment of the status of a sewer is required.

Definition of a public sewer

Two basic questions

17.   The approach to identifying a public sewer requires two separate questions to be answered, namely 'Is it a sewer?', followed by 'Is it a public sewer?' For the answer to the second question, a series of test questions should be applied in order to determine the answer, as detailed in the flow chart (Fig. 1).

What is a sewer?

18.   In cases of doubt, the question 'What is a sewer?' can frequently be answered by consideration of the following matters. According to S.343(1) PHA 1936 and S.219(l) WIA 1991, a 'drain' means a drain used for the drainage of one building or of any buildings or yards appurtenant to buildings within the same curtilage (the courts define 'curtilage' as being 'ground which is used for the comfortable enjoyment of a house and thereby as an integral part of the same, although it has not been marked off or enclosed in any way. It is enough that it serves the purposes of the house in some necessary or reasonably useful way'; *Sine lair-Lockhart's Trustees v. Central Land Board* (1950) 1 P&CR 195). A 'sewer' is not expressly defined, but the above sections state that the word 'sewer' 'does not include a drain as defined in this section, but includes all sewers and drains used for drainage of buildings

48

and yards appurtenant to buildings'. It can therefore include a 'single private drain' (Section 19 of the Public Health Acts Amendments Act 1890) or a 'combined drain' (S.38 PHA 1936), provided that in either case it serves as a sewer for buildings which are not within the same curtilage. A 'private sewer' is any sewer which is not a public sewer, since the terms 'sewer' and 'drain' are mutually exclusive.

Judicial interpretation

19.   *Open channels as sewers.* The status of a sewer does not depend on the nature of the effluent it carries. Accordingly, a channel may be a sewer although it carries only clean surface water (*Falconar* v. *South Shields Corporation* (1895) 11 TLR 223). This case demonstrated that the channel need not necessarily be of artificial construction in order to be a sewer, although a natural stream cannot normally be a sewer. There can therefore be such things as 'open surface water sewers'.

20.   It is a matter for question whether or not a land drain can ever be a sewer, but it would certainly be so only in unusual circumstances (see Garner and Bailey 1995, Ch. 1). It could be a sewer for the purpose of Section 4 of the Public Health Act 1875, but not a public sewer since land drains are expressly excepted (Section 13 of the Public Health Act 1875). Before 1875, some large land drains were referred to as 'sewers' (common usage), the *Shorter Oxford English Dictionary* defining a sewer as 'an artificial watercourse for draining marshy land and conveying surface water into a river or sea'.

21.   *Cesspools/septic tanks at end of sewer.* There are a number of common law cases which suggest that, if a conduit leads only to a pit or storage tanks, it is not a sewer (though for a different view of the case law, see Howarth and Brierley, 1995, p. 30). It is often said that a sewer must be in some form of a line of flow by which sewage or water of some kind should be taken from one point to another point, and there discharged by means of a 'proper outfall' (*Meader* v. *West Cowes Local Board* (1892) 3 Ch. 18). An example might be a line of pipes leading into a tank, from which there is a proper overflow and discharge arrangement, i.e. a 'proper outfall'. A line of pipes terminating in a cesspool, from which there is no 'proper outfall', is not, therefore, a sewer, nor is a conduit linking two cesspools. A 'proper outfall', according to common law precedent, does not mean that the outfall arrangements must be free from nuisance. In *Clark* v. *Epsom Rural District Council* (1929) 1 Ch. 287, the nuisance aspect was deemed not to prevent a line of pipes draining to a septic tank from being a

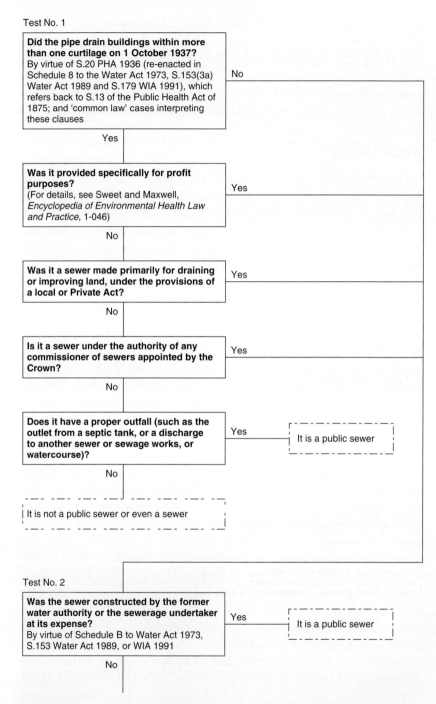

Test No. 1

Did the pipe drain buildings within more than one curtilage on 1 October 1937?
By virtue of S.20 PHA 1936 (re-enacted in Schedule 8 to the Water Act 1973, S.153(3a) Water Act 1989 and S.179 WIA 1991), which refers back to S.13 of the Public Health Act of 1875; and 'common law' cases interpreting these clauses

No

Yes

Was it provided specifically for profit purposes?
(For details, see Sweet and Maxwell, *Encyclopedia of Environmental Health Law and Practice*, 1-046)

Yes

No

Was it a sewer made primarily for draining or improving land, under the provisions of a local or Private Act?

Yes

No

Is it a sewer under the authority of any commissioner of sewers appointed by the Crown?

Yes

No

Does it have a proper outfall (such as the outlet from a septic tank, or a discharge to another sewer or sewage works, or watercourse)?

Yes

It is a public sewer

No

It is not a public sewer or even a sewer

Test No. 2

Was the sewer constructed by the former water authority or the sewerage undertaker at its expense?
By virtue of Schedule B to Water Act 1973, S.153 Water Act 1989, or WIA 1991

Yes

It is a public sewer

No

Fig. 1 (above and facing) Flow chart for identifying a public sewer

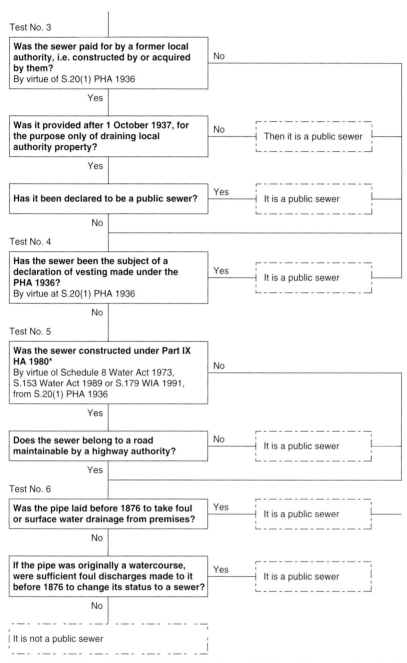

Test No. 3

Was the sewer paid for by a former local authority, i.e. constructed by or acquired by them?
By virtue of S.20(1) PHA 1936 — No

Yes

Was it provided after 1 October 1937, for the purpose only of draining local authority property? — No — Then it is a public sewer

Yes

Has it been declared to be a public sewer? — Yes — It is a public sewer

No

Test No. 4

Has the sewer been the subject of a declaration of vesting made under the PHA 1936?
By virtue at S.20{1) PHA 1936 — Yes — It is a public sewer

No

Test No. 5

Was the sewer constructed under Part IX HA 1980*
By virtue ol Schedule 8 Water Act 1973, S.153 Water Act 1989 or S.179 WIA 1991, from S.20(1) PHA 1936 — No

Yes

Does the sewer belong to a road maintainable by a highway authority? — No — It is a public sewer

Yes

Test No. 6

Was the pipe laid before 1876 to take foul or surface water drainage from premises? — Yes — It is a public sewer

No

If the pipe was originally a watercourse, were sufficient foul discharges made to it before 1876 to change its status to a sewer? — Yes — It is a public sewer

No

It is not a public sewer

*This superseded the provisions of Part IX of the Highways Act 1959, which may be relevant for the purpose of historical enquiry

Fig. 1 Continued

sewer, which was deemed, in that particular case, to vest in the local authority as a public sewer, under the provisions of the Public Health Act 1875 (see *Encyclopedia of Environmental Health Law and Practice*, 1-587).

22. *Pumping mains.* The definition of a sewer is not affected by the fact that the sewer contents may be pumped, as in the case of a pumping main (see 7.24 below).

23. *'Sewer' and 'drain': ordinary usage of terms.* In the absence of specific judicial guidance, the terms 'sewer' and 'drain' should be interpreted in accordance with the ordinary usage or dictionary meaning of the words. Thus, a pipe conveying surface water only from several houses to a soakaway would be considered as a 'drain', not a sewer (see also 7.20 above and S.219 WIA 1991), though the situation might be different where the system has been approved by the local authority (see *Attorney-General* v. *Peacock* (1926) 1 Ch. 241).

24. *Structures on a sewer.* S.219(2)(a) WIA 1991 states that references to a pipe, including references to a main, a drain or a sewer, shall include references to a tunnel or conduit which serves or is to serve as the pipe in question and to any accessories for the pipe.

25. *Sewer having several functions.* A pipe will be a public sewer if it serves the defined functions of a public sewer, even though it serves other purposes as well (see *Encyclopedia of Environmental Health Law and Practice*, 1–069 and 1–587; also *Hutton* v. *Esher Urban District Council* (1974) Ch. 167).

Tests to check the status of a public sewer

26. The tests in the flow chart (Fig. 1) are not intended to be an exhaustive guide, particularly in difficult cases, for which legal advice should be sought. Nevertheless, a ready solution to most of the types of situation encountered by the practising engineer will be obtained by applying the 'question and answer' route embodied in the flow chart.

Future legislation

27. From 2011, all private sewers and lateral drains that drain to the public sewerage network of the nine English water companies will transfer to water company ownership. This is intended to simplify the responsibilities for an integrated sewerage network and improve long-term planning, which will allow more efficient adaptation to the effects of climate change and housing growth. This followed a consultation in

July 2007, held jointly with the Welsh Assembly Government (WAG). The WAG published a Strategic Policy Statement on Water on 31 March 2009, which outlines how the transfer will be implemented in Wales. The UK Government intends to consult on new draft regulations for implementation in England later in 2009.

8

Development either side of a watercourse

1.   The following observations may be of assistance to engineers who encounter a situation where development on either side of a watercourse gives rise to urban land drainage problems.

2.   Houses have often been built on either side of a watercourse with their gardens extending to it. Alternatively, a watercourse may pass the frontage of the properties, between them and the highway. In both cases, drainage problems may arise and require expenditure to put them right, and give rise to disputes over ownership (see Chapter 6 above).

3.   Section 62 of the Law of Property Act 1925 provides that 'a conveyance of land shall be deemed to include and shall by virtue of this Act operate to convey with the land all buildings, hedges, ditches, fences, ways, waters, watercourses, liberties, privileges, easements, rights and advantages whatsoever appertaining or reputed to appertain to the land or any part thereof.' This section operates only in the absence of any contrary intention expressed in the conveyance.

4.   There is a rebuttable presumption that where there is a ditch and a bank on the boundary of a property, the person who dug the ditch must have dug it at the extremity of his land and thrown the soil on his own land to make the bank. If there is evidence to identify a different boundary, this presumption will be rebutted (see *Fisher* v. *Winch* (1939) 2 All ER 144).

5.   In general, the ownership, control and occupation of the bed or banks of a watercourse are even more relevant to land drainage problems than are riparian rights. It may be seen from the discussion above, therefore, that where the householder's own land adjoins a channel, his responsibilities are not always absolutely clear. It is sometimes found that the deeds to the properties extend to the boundary of the watercourse, and sometimes to a fence or hedge set on the bank of the watercourse.

6. Nevertheless, the wording of the deed may not be conclusive because of the (rebuttable) legal presumptions in Section 62 of the Law of Property Act 1925, and that a person whose land abuts on a river owns the bed of the river up to the middle (*Blount* v. *Layard* (1891) 2 Ch. 681).

7. A drainage board or local authority that wishes to serve notice under S.25 LDA 1991, requiring works to maintain the flow of a watercourse, may encounter an inability or unwillingness of the owners to accept that they are responsible for paying for, or carrying out, desirable, although possibly expensive, works. It will help to minimise legal confusion over establishing responsibilities if officers familiarise themselves with the law in such cases.

8. District councils and unitary authorities have a discretionary power to carry out works themselves, for the prevention of flooding, in this type of case (SS.14 and 15 LDA 1991). If the authority decides to do so, it will then generally wish to accept responsibility for subsequent maintenance, in order to prevent future problems. In many cases, therefore, the problem may ultimately relate to an inability or unwillingness to pay, rather than to a real problem of confusion on the legal question of responsibilities.

9. S.60 LDA 1991 provides that a local authority may contribute to the expenses of the execution or maintenance of any drainage works by a drainage body 'such an amount as, having regard to the public benefit to be derived there from, appears to the local authority to be proper'. No power exists to contribute towards the cost of an individual's private scheme. S.60 LDA 1991 is, in fact, used to empower contributions between local authorities and other drainage bodies.

10. When development takes place on land adjoining a main river or ordinary watercourse, it has become common practice for the planning authorities (on the advice of the Agency or an IDB) to require the developer to leave a strip of land, at least 5–8 m wide, free of development along one or both sides of a watercourse, in order to provide access for future maintenance. This strip also provides a suitable area for spreading spoil, but problems can be created unless careful consideration is given to detail. Under most Agency Regional Land Drainage Byelaws, these strips can be required under the formal consenting procedure for main river watercourses, and this requirement is not excluded by the granting of any planning permission under the TCPA 1990 (see Chapter 10 below). A waste management licence for such spreading is not required if the deposits result in agricultural or ecological improvements (Waste Management Licensing Regulations

1994, Regulation 17 and Schedule 3, paragraph 25; Department of Environment (DoE) Circular 11/94, paragraph 5.74). Other waste management regulations which may apply are the EU Waste Directive 2006/12/EC, EU Landfill Directive 99/31/EC, Landfill (England and Wales) (Amendment) Regulations 2004 and Hazardous Waste (England and Wales) (Amendment) Regulations 2009.

11.   In promoting its flood defence and other duties, particularly those involved with conservation and the environment, the Agency regards the land associated with a watercourse as a 'river corridor' to be protected and enhanced. Use of this corridor will include maintenance purposes, although this is only one of a whole range of water environment interests.

12.   Similarly, some river corridors of particular nature conservation importance are now being designated as Sites of Special Scientific Interest, under the Wildlife and Countryside Act 1981 (see further in 17.11–17.13 below). In preparing development plans for their areas, planning authorities must now have regard to linear and continuous structures such as rivers and their banks, where these are essential for 'the migration, dispersal and genetic exchange of wild species' (Regulation 37 of the Conservation (Natural Habitats, etc.) Regulations 1994; see 10.11 below).

13.   The width and configuration of any corridor should be the subject of joint agreement between the drainage body and/or the Agency and the local planning authority, depending on particular circumstances. Environmental and topographic surveys, plus other studies, are now frequently required in order to determine the desirable corridor extent and configuration.

14.   Development up to the bank of any watercourse should be avoided. It may amount to an assertion of land ownership, but will create problems of access, damage to the banks, risk of disposal of garden refuse into the watercourse, and generally be contrary to the need to preserve an attractive river corridor.

15.   The careful choice of layout of any new development should include the enhancement of the watercourse as a feature, perhaps with houses, footpaths and roads built to face it, and to have the corridor identified as a 'green chain' of open spaces.

16.   This enlightened approach to the treatment of river corridors benefits from co-ordinated planning at an early stage in the process. To avoid the problem of unsightly and neglected land, where possible the ownership should not be left with the developers but preferably with a responsible body or organisation with such agreements as may be appropriate to

ensure the aftercare and maintenance of the land. Commuted sums of money could possibly be obtained from the developer for this purpose (see 10.27–10.28 below).

17.   It may be possible for the open land to be maintained by the local authority as public open space and amenity land, using powers under Section 137 of the Local Government Act 1972.

18.   Small sites and redevelopments can have particular problems, and the local authority may seek other informal options, partly because of the expenses involved. It may, for example, request the developer to include a restrictive covenant within the deeds of the properties, although it may not be able to insist on this as a planning condition. In cases where no maintenance strips exist to provide access to the watercourse, the powers available to drainage bodies under SS.165 and 167 WRA 1991 and SS.14, 15 and 64 LDA 1991 are adequate for the purpose of carrying out maintenance works. However, considerable inconvenience, including practical problems and public relations difficulties, may arise in the exercise of those powers. High compensation costs may have to be paid, especially if there is disruption of a number of well-established gardens.

19.   Some statutory control of the provision of maintenance strips is available to the Agency under its Regional Land Drainage Byelaws for main rivers (see 8.10 above). Similarly, the establishment by IDBs, district councils or unitary authorities of byelaws for ordinary watercourses is possible under S.66 LDA 1991, and is thus to be encouraged to improve control. Byelaws might require consent to be obtained before any structures, etc., are erected, or land filling is carried out within a defined distance from a watercourse. Significantly, those works of drainage bodies or the Agency exempt from a planning consent under the provisions of the Town and Country Planning (General Permitted Development) Order 1995 (Schedule 2, Parts 14 and 15), but for which an environmental assessment would not be required, may thereby be brought under control (see generally Chapter 10).

9

Extension of the 'main river' designation of watercourses

1.   As noted previously, a crucial distinction in land drainage and flood defence law is that between main rivers and ordinary watercourses (see 1.9 and 3.27–3.28 above). Practical problems have been encountered with regard to the designation of further stretches of watercourses as main rivers, but the legal position in this matter is reasonably clear.

2.   Some of the difficulties reported concern situations in which a local authority is prepared to improve a watercourse provided that the Agency agrees to recommend its designation as a main river thereafter. However, such agreements may involve what are seen as expensive design requirements being specified by the Agency. Whether or not the suggested design requirements are actually excessive is a matter for agreement between the engineers of the respective authorities. An example might be a flood alleviation scheme carried out under the powers conferred by SS.14–16 LDA 1991, where the consent of the Agency is required (see 3.9 above).

3.   In such cases, it is highly desirable that the design parameters and methods should be discussed and established at an early stage. Where grant aid is being obtained from the Agency, early agreement should also be sought with Defra (for addresses, see Appendix 3 below). Funding allocation is discussed further in Chapter 14 below.

4.   If alternative design parameters and costings can be established and discussed at a sufficiently early stage, this can sometimes avoid arguments later from more entrenched positions. The statutory duty of the Agency, however, is to exercise overall supervision and to provide the lead in making decisions concerning standards (S.6(4) EA 1995).

5.   A number of engineers have suggested the possibility of extending main rivers upstream, either to the uppermost outfall of a sewage works or to a public sewer outfall on the watercourse. However, it is

for the RFDC to recommend the extent of the main river designation, taking due consideration of the resource implications of the designation.

6.   There are, however, a considerable number of small works and public surface water sewers that discharge into watercourses that are well upstream of the present limits of main rivers. These watercourses are at present the responsibility of the riparian owners, and no cost falls on the ratepayers. Local authorities, however, possess the discretionary power to carry out improvements in relation to flood alleviation, e.g. the powers conferred by SS.14–16 LDA 1991 referred to above. When a proposal to extend a main river is being considered, the Agency may require the prior improvement of the watercourse.

7.   It should not be overlooked that the decision to designate a watercourse as a main river rests with the appropriate Secretary of State rather than with the Agency or RFDC, and that consultation with the Secretary of State in particularly difficult cases may enable a solution to be agreed between all concerned. This type of problem appears, therefore, to be capable of resolution, given a reasonable and co-operative approach by the parties.

10

Development planning and management

Planning procedures

1. Obtaining the legal right to develop is usually granted either by the local planning authority or under permitted development rights, and will be a key stage in every development with land drainage implications. However, it should be stressed that obtaining planning permission does not avoid the need to obtain other authorisations where these are required, for example from local authorities in relation to culverting.

2. In respect of proposals for development involving land drainage, a developer may submit either an 'outline' or a 'detailed' application for planning permission to the local planning authority, which exercises powers under the provisions of the TCPA 1990. The local planning authority which determines applications for planning permission is either the district council or relevant unitary authority. By virtue of Article 3 of the GDPO, outline applications can be made only for the erection of buildings.

3. Planning permission will be required for any development consisting of a 'building, mining, engineering or other operation' or for a material change in the use of land. Most drainage works are likely to be 'engineering or other operations'. However, certain kinds of development which would otherwise require planning permission are 'permitted development' in planning law for which planning permission is deemed to be granted. The most important kinds of permitted development in the present context are those contained in Schedule 2 of the GDPO, namely:

    (a) Part 6, Class A (engineering operations reasonably necessary for agricultural purposes on agricultural units over 5 hectares)
    (b) Part 14 (development by drainage bodies)

(c) Part 15 (development by the Agency)

(d) Part 16 (development by or on behalf of sewerage undertakers).

4. An outline application seeks to establish in principle whether a proposed building, including the method of drainage, may be adequately provided on the site. PPS25 draws attention to the need to take land drainage as part of a comprehensive approach to the assessment of flood risk from all sources when planning applications are being determined.

5. In all cases, when submitting planning applications, whether in full or outline, there has been, since April 2008, standard national requirements for the validation of planning applications under Department of Communities and Local Government (DCLG) Circular 02/2008. This introduced standard application forms across the country and the power for local authorities to publish their own Local Validation Requirements list. The local authority should always be contacted to ensure that all requirements in terms of land drainage, flooding and other matters have been covered prior to submitting and application. Unless the local planning authority is satisfied that all information provided in Annex A of the circular and its local requirements have been satisfied, it can refuse to validate the application. Its requirements should, of course, be proportionate to the nature of the scheme, and not all of the information on the local authority's published list will be necessary in all cases.[1]

6. If outline permission is given, then certain 'reserved matters' may be identified. Reserved matters are defined in Article 1 of the GDPO as matters specified by condition for which subsequent approval is required. For applications relating to sites not previously developed, detailed drainage considerations are frequently reserved, provided that the general principles relating to drainage have been established.

7. Following the coming into force, on 1 October 2006, of the amendment to Article 10 of the GDPO, local planning authorities are required to consult the Agency as a statutory consultee on the following types of application:

Development involving:

(a) the carrying out of works or operations in the bed of, or within 20 metres of the top of a bank of, a main river which has been

---

[1] There are provisions for appeals against non-validation of planning applications under S.78 TCPA 1990, based on a ground of non-determination within the 8- or 13-week determination period.

notified to the local planning authority by the Agency as a main river for the purposes of this provision or

(b) the culverting or control of flow of any river or stream.

Development, other than minor development, which is to be carried out on land:

(a) in an area within Flood Zones 2 or 3 or

(b) in an area within Flood Zone 1 which has critical drainage problems and which has been notified for the purpose of this provision to the local planning authority by the Agency.

Any development of land of one hectare or more.

8.   If an objection is received from the Agency in respect of an application on the grounds of flood risk, and the local planning authority is minded to approve such an application, this must be referred to the Secretary of State to determine whether the application can be determined by the local planning authority or whether it is to be called in for determination by the Secretary of State (DCLG Circular 02/2009).

9.   The importance is stressed of full consultation at outline and even pre-submission stages, in order to establish drainage principles as early as possible and to ensure that all validation requirements for the planning application are agreed. In some situations, the Agency may transmit its observations not only to the planning authority but also to the applicant directly.

10.   Where the planning application relates to a Site of Special Scientific Interest under Section 28 of the Wildlife and Countryside Act 1981 (WCA 1981) (see also 17.13–17.15 below), or an area within 2 kilometres of such a site which has been notified to the planning authority by the relevant nature conservation agency (Natural England or the Countryside Council for Wales (CCW)), then that conservation agency will also be a statutory consultee in the planning process (Article 10 of the GDPO).

11.   In the exercise of its powers, the local planning authority must satisfy itself as to the suitability and feasibility of a proposal. In doing so, it must take into account any consideration (e.g. land drainage arrangements) which is material, but it must take as the starting point for any decision the provisions of the relevant development plan (Section 38(6) of the Planning and Compulsory Purchase Act 2004 (PCPA 2004)), which states that 'If regard is to be had to the development plan for the purpose of any determination to be made

under the planning Acts the determination must be made in accordance with the plan unless material considerations indicate otherwise.' The PCPA 2004 requires local planning authorities to contribute to the achievement of sustainable development, and, as part of that, they are required to give consideration of environmental impact in the decision-making equation. Government policy guidance plays a central role in determining the weight to be given to environmental considerations, as against other material considerations. Planning Policy Statement 1, *Delivering Sustainable Development*, is also a key cornerstone of the planning system, seeking to encourage a proactive and positive role for planning in improving the quality or development and protecting the environment.

12.   Key policy guidance is listed in Appendix 1 below.

13.   The system of development plans in operation within England and Wales was also revised under the PCPA 2004. Replacing the former system of the Structure Plan made by the county council and the Local Plan made by the district council. In areas where there is a unitary authority, this also replaces the Unitary Development Plan, comprising both strategic and local matters, though transitional arrangements will apply where a unitary authority has been newly established. This new system, known as the Local Development Framework, is effectively a suite of documents which cover matters such as strategic policies (contained within the 'core strategy' and more specific matters such as development control policies and policies which allocate land for development). Each authority will be at differing stages of plan preparation, and contact with them is advisable to establish the current status of different parts of the development plan. Many authorities still have a number of old style development plan policies which they have been permitted by their relevant government office to continue using ('saved').

14.   Given the increasing importance accorded to development plans in the planning process, the importance of interested parties influencing the shape and provisions of development plans cannot be understated. As with planning applications, there are detailed rules relating to consultation and the making of representations. Under these, a variety of organisations, including the Agency, other local authorities and the statutory conservation and countryside agencies, must be consulted during the process of preparing the plan. Organisations are strongly encouraged to engage with the local planning authority as early as possible in the plan-making process, making it easier for comments and concerns to be taken on board before the plan is finalised. There

are still formal opportunities to be consulted on the public consultation draft, prior to the plan being submitted to an independent planning inspector for scrutiny (S.20 PCPA 2004). In recent years the importance of addressing environmental impact in development plans has been emphasised. This is set out in S.39 PCPA 2004. However, the former Office of the Deputy Prime Minister issued in November 2005 *Sustainability Appraisal of Regional Spatial Strategies and Local Development Documents* (see also Planning Policy Statement 12: *Local Spatial Planning*). Attention is also drawn to the need for habitat conservation to be addressed in development plans. This has become more formalised with the implementation of the Habitats Regulations, which requires an assessment of the direct and indirect impact of development plans on European Protected Sites (referred to as 'appropriate assessment'). This requirement is outlined in Article 6(3) and (4) of the European Communities (1992) Council Directive 92/43/EEC on the conservation of natural habitats and of wild fauna and flora (the 'Habitats Directive').

15.   Since the Agency has only limited statutory powers in connection with the control of surface water drainage works, it is essential for a strong partnership to exist with the local planning authority. This will ensure that adequate constraints exist, and, in particular, will encourage opportunities for enhancements which take the full range of environmental factors into account. PPS25 and an associated 'living draft' practice guide (published in February 2007) set out national policy requirements in relation to the management of all sources of flooding. Guidance on legislation on this matter is evolving rapidly. Indeed, The Government published the Flood and Water Management Bill in April 2009. This aims (amongst other provisions) to make clear who is responsible for managing all sources of flood risk and to encourage more sustainable forms of drainage in new developments (Defra).

16.   Attention is also drawn to the Construction Industry Research and Information Association's (CIRIA) Report Nos. 123 and 124 (*Scope for Control of Urban Runoff*, Volumes 1 and 2), Project No. 448 (*Manual on Infiltration Methods for Storm Water Runoff Control*) and Report No. 142 (*Control of Pollution from Highway Drainage Discharges*). Indeed, encouragement should be given to such drainage methods as soakaways, surface swales and enhanced methods of infiltration and attenuation in appropriate locations.

17.   Co-ordinated drainage schemes, integrated with landscaping, roads, and infrastructure, can be implemented successfully only if the principles are established at the pre-planning or outline stage.

Consideration of land drainage matters

18.   PPS25 draws attention to the need to ensure that where land drainage considerations arise, they are always taken into account in the determination of planning applications, as part of the comprehensive assessment of the risk of flooding from all sources. Developments approved without regard to land drainage problems can endanger life, and result in damage to property and wasteful expenditure of public resources on remedial works, whether on the development site or elsewhere. For this reason, planning permission may be refused where unsatisfactory provision for drainage has been made (*George Wimpey and Co. Ltd* v. *Secretary of State for the Environment* (1978) JPL 776; other related planning decisions are referred to in Howarth and Brierley, 1995, p. 10). It is important for planning authorities to consult the Agency where land drainage considerations arise and for subsequent close liaison between the planning authority, local authority engineers and the Agency. Particular attention should be paid to the desirability of consultation where development is proposed on a floodplain or wash land, or in an area which the Agency has indicated might be subject to drainage problems or be susceptible to inundation by the sea or tidal flooding.

19.   The importance of giving careful consideration to these matters at the planning application stage cannot be overemphasised, and such an approach forms the foundation of the 'prevention rather than cure' goal of the drainage engineer.

20.   The use of planning conditions provides a vital controlling mechanism which will minimise future drainage problems consequent upon new development (see 10.22–10.28 below). Additionally, the local planning authority is able to ensure that planning conditions are complied by the use of the range of enforcement powers provided under Part VII of the TCPA 1990.

21.   Where development takes place in the area of an adjacent local planning authority, problems should be avoidable, as the Agency must be consulted by the authority and may comment on any land drainage requirements needed to deal with the increased runoff. As already referred to in 10.6 above, the Agency is a statutory consultee, and should the local planning authority wish to approve a scheme despite such an objection, the matter must be referred to the Secretary of State.

Conditions for consent

22.   The scope of planning conditions is important when problem areas which can arise in connection with land drainage are being considered.

23.   S.72 TCPA 1990 makes reference to the use of conditions for 'regulating the development or use of land under the control of the applicant (whether or not it is land in respect of which the application is made) or requiring the carrying out of works on any such land so far as it appears to the local planning authority to be expedient for the purposes of or in connection with the development authorised by the permission'. The extent to which conditions may be imposed is now dealt with in DoE Circular 11/95, *The Use of Conditions in Planning Permissions.*

24.   Planning authorities have a wide area of discretion, although there is a presumption against using town planning controls in situations where powers exist elsewhere within other legislation (see DoE Circular 11/95, paragraph 22). It is also accepted that conditions must serve some genuine planning purpose in relation to the development permitted, and must not be used to achieve a totally extraneous objective.

25.   Thus, in *Pyx Granite Co. Ltd* v. *Ministry of Housing and Local Government* (1958) 1 QB 554, Lord Denning remarked that, 'although the planning authorities are given very wide powers to impose such conditions as they think fit nevertheless the law says that these conditions, to be valid, must fairly and reasonably relate to the permitted development. The planning authority are not at liberty to use their powers for an ulterior object, however desirable that object may seem to them to be in the public interest.' The courts have also held that conditions should not render development incapable of being carried out.

26.   It will be appreciated from the above that a condition cannot be imposed that requires the developer to carry out drainage works downstream on land which is not owned by the developer, if the owner of such land is not prepared to allow the works to proceed. Nor is it possible for a local authority to use the powers relating to land drainage or public health to require downstream owners to improve the existing land drainage system.

27.   Although it cannot be made a condition of a planning consent that a developer must carry out works on land not under his control, consent can nevertheless be refused unless he agrees to one of the options outlined in 10.34 below.

28.   Control over the early provision of drainage infrastructure to serve a new development where it is either within the developer's control or within a public highway adjacent to the site may, however, be provided by a negatively phrased condition. A condition requiring the provision of such drainage facilities before the commencement of the development of the site is a useful way of ensuring that, for example, vulnerable

land adjacent to a site is not flooded during the development process (see *Grampian Regional Council* v. *City of Aberdeen District Council* (1984) 47 P&CR 633; and Planning Policy Guidance 13, *Transport*, Annex C, and its Welsh equivalent (Planning Policy Guidance 13, *Highway Considerations in Development Control*)).

Planning obligations

29.   Under S.106 TCPA 1990 (formerly Section of the 52 Town and Country Planning Act 1971; and now amended by Section 12 of the Planning and Compensation Act 1991), the local planning authority may enter into an obligation with any person interested in land in its area for the purpose of restricting or requiring specified development or use of the land, and in order to incorporate any provisions (including financial ones) 'as appear to the local planning authority to be necessary or expedient'. Such obligations comprise both planning agreements and unilateral undertakings by developers (see also 5.30 above).

30.   Government Circulars have given encouragement to the use of such obligations for the phasing of drainage provisions, but have emphasised that obligations should be sought only where necessary to the granting of permission, relevant to planning and relevant to the development to be permitted (Circular 05/05, *Planning Obligations*). However, the courts have held that the 'necessity' test is not a strict one. As long as a planning obligation has some connection with a proposed development, it may be a material consideration which the planning authority must have regard to, though not necessarily accept. Within reason, it is up to the planning authority to decide how much weight to place on the obligation (*Tesco Stores Ltd* v. *Secretary of State for the Environment* (1995) 2 All ER 636).

Requirement for information and advice

31.   As stated in PPS25, an applicant for planning permission should provide any information required to confirm that the drainage arrangements are satisfactory, and the planning authority, usually advised by drainage engineers, will require the applicant to supply any additional information in order to enable it to consider the application.

32.   The drainage authority, whether the Agency or local authority engineers, should give advice which includes an assessment of the flooding effect downstream, and should make suggestions as to what drainage works, if any, would prevent it. Where available, surveys carried out

under S.105(2) WRA 1991 will provide useful information for this purpose (see 3.13–3.14 above). With the increasing acceptance and promotion of Catchment Flood Management Plans (CFMPs) principles by the Agency, local planning authorities and the statutory nature conservation agencies (see also 10.43 below) it may be necessary for the developer to undertake or finance surveys and studies to investigate the drainage mechanisms. Identification of possible options and solutions to minimise the impact on watercourses and river corridors will also be required.

33.   Environmental Impact Assessments (EIAs), including the use of baseline studies, are now increasingly being recommended by the Agency as part of its duty to promote conservation and enhancement of the river environment. Where certain projects are likely to have a significant effect upon the environment, there is a legal requirement for the developer to submit a formal Environmental Statement (ES) as required under either European or national law (see Chapter 17 below).

Requirements for works, agreements, etc.

34.   Where it appears to the planning authority (acting on representations from local authority engineers or the Agency) that downstream works to a watercourse are required in order to cater for a proposed development, the developer may be advised as to the matters listed below.

(a) A downstream length of a watercourse may be included in an amended application, which would include particulars and a plan of the drainage work. Where an application is extended to include drainage works, the planning authority may incorporate in the permission a condition requiring these works to be carried out first (but see 10.23–10.28 above).

(b) A formal agreement may be entered into between the developer and the owners of the land downstream, requiring the applicant to carry out all necessary works, or for the downstream landowners to carry out and maintain the necessary works in return for a payment by the developer. Under this type of agreement, the body or individual responsible for enforcement should be identified, as the council or Agency may not be a party.

(c) On-site works may be provided for infiltration and source control and/or storage and controlled release of surface water, which would render the downstream works unnecessary. Consideration should be given at this stage to the responsibility for future maintenance of such on-site drainage features. If they form part of a highway drainage system, they will frequently be adopted by the highway authority by an agreement under S.38 HA 1980. If they form part of a surface water sewerage system, they will normally be adopted by the water company by agreement under S.104 WIA 1991.

(d) A planning obligation under S.106 TCPA 1990 may be entered into with the local authority, incorporating provisions concerning development or infrastructure works which have some connection with the proposed development (see 10.29–10.30 above).

(e) A sewer (as an alternative to improving a watercourse) may be requisitioned under the provisions of S.98 WIA 1991, where drainage of domestic premises is involved.

35.   In cases (a) to (d), the future maintenance arrangements for the required drainage works should be carefully considered, bearing in mind that developers move on, and a variety of subsequent purchasers and owners downstream may often not be willing to execute maintenance obligations.

Further practical considerations

36.   Despite potential drainage problems caused by new development being highlighted by the now superseded Ministry of Housing and Local Government Circular 94/69 as a matter which demanded rigorous investigation at the planning stage, many unsatisfactory developments have occurred since that time, and drainage problems have resulted. This is also despite the various preventive measures described in 10.22–10.34 above being at the disposal of those considering the application.

37.   Awareness by drainage authorities of a potentially problematic development is the first requisite of a sound approach to such matters. Local authority engineers can play a vital role in alerting planning officers to those proposals where land drainage considerations are likely to be of major importance. Scrutiny of incoming planning applications by a local authority drainage engineer, with his knowledge of local flooding problems, ground permeability characteristics and the existing drainage network, can be invaluable at this early stage when drainage principles need to be established and potential problems highlighted. Close liaison between local authority planners and engineers is essential if this process is to work effectively.

38.   Close liaison between the local authority and the Agency is also important throughout the processing of planning applications. Early

notification by the Agency of those applications it intends to object to, or to make significant representations about, is often of value in order that applications are not processed before such responses are received.

39.   On receipt of representations from the Agency (and others), the local authority engineer may often be required to advise the planning officer on the most effective course of action to follow regarding drainage considerations, given his own local expertise, and what are often less site-specific observations. Local planning authorities must accept their role of monitoring and advising on surface water drainage works on watercourses other than main rivers. The commitment and resources should be made available (perhaps by using consultants) to ensure that adequate assessments are made at the earliest possible opportunity.

40.   Some local planning authorities are under pressure to leave all comments on land drainage matters to the Agency rather than treating the achievement of satisfactory control as a joint responsibility. Were they to do this, though, they would be in breach of their duty to determine planning applications according to *their* assessment of the strength to place on all the material considerations. However, the general situation is improving, with more enlightened attitudes beginning to gain wider acceptance.

41.   Following agreement with the developer for the provision of the type of works outlined in 10.29–10.30 above, or the imposition of conditions relating to drainage matters, the future monitoring of the works is as important as the initial negotiation of appropriate design details. Again, the local authority engineer is best placed to ensure that such works are undertaken, especially through experience of S.104 WIA 1991 and S.38 HA 1980 adoption works, and contact with building inspectors, who may receive drainage proposals concurrently.

42.   In short, the key to the effective prevention of potential drainage problems and damage to the environment caused by new development is: firstly, the early recognition of problematic proposals; secondly, the rapid notification of the planning officers of such; thirdly, the close liaison between planners, drainage engineers, developers and Agency staff to determine the best means of preventing the potential problem; and fourthly, the monitoring of the implementation of the agreed details.

43.   Since 1974, the concept of integrated river basin management has become firmly established; with the creation of the NRA, further opportunities arose for this concept to be built upon and broadened into Catchment Management Plans (CMPs), and it is likely that this approach will be actively continued under the Agency. The process

involves the identification of problems and constraints to development in the catchment of a watercourse on matters such as flood defence, water resources, water quality, fisheries, navigation, recreation, and conservation of the environment. Since 1974 the concept of integrated river basin management was firmly established and is now being managed by the Agency with Catchment Flood Management Plans (CFMPs). These are a planning tool through which the Agency agrees long-term sustainable policies for flood risk management in partnership with other key decision-makers within a river catchment. CFMPs are a learning process to support an integrated approach to land use planning and management, and also River Basin Management Plans under the Water Framework Directive.

44.   An expansion of this baseline information will permit improved guidance and advice to be given to developers and local authorities on constraints and opportunities to assist with the strategic planning of sustainable development. The incorporation of appropriate policies is beginning to have a significant influence and is seen as part of the CMP process. Traditional land drainage advice now has to expand to accommodate flood defence and the broader issues of environmental protection and enhancement which the Agency has a duty to promote.

11

Development in floodplains

1. There is greatly improved direct control of developments in flood-plains from the Agency's point of view, following PPS25, which came into force in December 2006. The principal executive power still lies with the local planning authority's response to representations made by the Agency (and others) and other material considerations, but there is a clearer framework for deciding what should be permitted development. This is explained below.

2. In relation to a main river, the generally accepted view of the Agency is that the powers provided by S.109 WRA 1991 (see 3.23 above) relate to the control of works within a main river, i.e. between the river banks. However, if artificial flood banks exist or are constructed, then the powers are interpreted to include both them and the intervening land in the definition for control purposes.

3. Alongside the powers exercisable in relation to a main river, additional control is exercised by the Agency over structures, filling and earthworks in the floodplain under the provisions of each region's land drainage byelaws. However, other than within the maintenance strip defined in byelaws (generally 5 or 8 metres each side of the watercourse), the granting of a planning permission will, under usual byelaw provisions, exempt the works from the Agency's consent procedure (though it should be noted that there are exceptions to this, e.g. in the Agency's Anglian region, where any such heaps of materials in 'wash lands' outside of the main-tenance strip will still require specific consent under the byelaws). The Agency is therefore a statutory consultee under the planning process.

4. In this context, a 'permitted development' under the provisions of the GPDO is not exempted from the control of the byelaws. Only planning permission granted by the local planning authority or by the Secretary of State on an application in that behalf made to a local planning authority will remove the need for consent under the byelaws, and this will be following consultation with the Agency and preparation of an appropriate flood risk assessment by the developer.

5.   Agency advice to local planning authorities takes into account its responsibility under S.6(4) EA 1995 to exercise a general supervision over all matters relating to land drainage along with its general environmental and recreation duties under SS.6 and 7 EA 1995. General guidance for local planning authorities in relation to consultation with the Agency is contained in PPS25 (see also Chapter 10 above).

6.   Although it is generally accepted that under certain conditions, development can be accommodated in floodplain areas, the circumstances will involve the examination of strategic flooding issues, works of mitigation, and environmental conservation and enhancement. It should be appreciated that successive developments constructed without consideration of these factors could necessitate expensive remedial works being required to restore the protection of adjacent properties upstream (and possibly downstream) and to counteract the cumulative hazard caused.

7.   The Agency, although exercising regulatory powers, is not always opposed to new development or redevelopment, but advocates that such developments should be executed in accordance with accepted plans and policies.

8.   It must be recognised that it is generally not possible for a flood protection scheme to give absolute protection against flooding. A flood alleviation scheme will afford protection only within a fixed design limit, generally expressed in terms of the statistical likelihood of a flooding occurrence. The effectiveness of any mitigation measures must also be considered.

9.   A wide range of problems within the floodplain and river corridor can be created by mineral extraction site restorations and landfill waste disposal proposals. Landscaping, stockpiling and domed restorations can affect floodwater conveyance and storage areas. Groundwater levels, flows and quality can be affected by the drainage and infilling of pits. Low-level restorations and the resulting need for alternative drainage schemes require detailed consideration well in advance of any planning application. It is essential to encourage early consultation with the Agency on these issues if delay at the planning application stage is to be avoided. Hydrological, hydraulic and environmental studies are now frequently required by the Agency to satisfy it of the suitability of new schemes.

10.   Embankments for roads or railways across floodplains also require substantial investigations to be carried out before the Agency's support can be obtained. Works of mitigation to provide replacement of lost conveyance and storage facilities together with preservation and enhancement of the environmental aspects have to be considered.

Planning Policy Statement 25: *Development and Flood Risk* (PPS25)

11. PPS25 replaced DoE Circular 30/92, and aims to ensure that flood risk is taken into account at all stages in the planning process to avoid inappropriate development in areas at risk of flooding. It takes a risk-based approach to ensure development is safe where it is demonstrated to be necessary. There are a number of key planning objectives, as outlined below.

12. The aim of PPS25 is to ensure that flood risk is taken into account at all stages in the planning process, to avoid inappropriate development in areas at risk of flooding, and to direct development away from areas at highest risk. It also encourages a partnership approach between stakeholders.

13. Where new development is, exceptionally, necessary in high-risk areas, policy aims to make it safe without increasing flood risk elsewhere and, where possible, reducing flood risk overall.

14. PPS25 recommends that regional planning bodies and local planning authorities should prepare and implement planning strategies that help to deliver sustainable development by:

(a) Appraising risk – identify land at risk and the degree of risk of flooding from river, sea and other sources in their areas; and preparing RFRAs or Strategic Flood Risk Assessments (SFRAs).

(b) Managing risk – setting policies, e.g. local plans which avoid flood risk to people and property where possible, and managing any residual risk, taking account of the impacts of climate change; and only permitting beneficial development in areas of flood risk when there are no reasonably available sites in areas of lower flood risk.

(c) Reducing risk – safeguarding land from development that is required for current and future flood management, reducing flood risk for new development through location, layout and design, incorporating SUDSs; surface water management plans; flood storage, conveyance and SUDSs; recreating functional floodplain; and setting back defences.

Risk-based approach

15. A risk-based approach should be adopted at all levels of planning. A strategic approach should be taken when planning areas of development, such as avoiding inappropriate development in flood risk areas.

The design of a new scheme should look at managing flood pathways and the susceptibility to flooding.

16. A flood risk assessment should be carried out to the appropriate degree at all levels of the planning process, and for all forms of flooding to and from the development. Forms of flooding to be considered include coastal, fluvial, overland flows, groundwater, sewerage and other artificial sources (e.g. reservoirs and canals). CIRIA Report No. 624, *Development and Flood Risk – Guidance for the Construction Industry*, provides information on good practice in the assessment and management of flood risk as part of the development process. Part C of CIRIA 624 gives guidance on three levels of flood risk assessment, and which is appropriate to the development. Annex E of PPS25 gives the minimum requirements for a flood risk assessment. Table D3 of PPS25 compares flood risk and vulnerability and determines whether or not development is typically appropriate.

Sequential test

17. Local planning authorities allocating land for development should apply a sequential test to demonstrate that development has been steered to areas at the lowest probability of flooding, where appropriate. The flood zones are the starting point for this approach, and this information can be found on the Agency's flood maps, further supplemented by SFRAs, where available. In areas at risk of river or sea flooding, preference should be given to locating new development in Flood Zone 1. If no reasonably available sites exist in Zone 1, the flood vulnerability of the development can be taken into account (dependent on the proposed use, as defined in Table D.2 of PPS25) in locating development in Flood Zone 2, then Flood Zone 3.

18. Flood zones are defined in Annex D of PPS25, along with requirements for flood risk assessment. These zones are defined:

- *Zone 1*. Low probability – comprises land assessed as having a less than 1 in 1000 annual probability of river or sea flooding in any year ($>0.1\%$).
- *Zone 2*. Medium probability – comprises land assessed as having between a 1 in 100 and 1 in 1000 annual probability of river flooding ($1–0.1\%$) or between a 1 in 200 and 1 in 1000 annual probability of sea flooding ($0.5–0.1\%$) in any year.
- *Zone 3a*. High probability – comprises land assessed as having 1 in 100 greater annual probability of river flooding ($>1\%$) or a 1 in 200 or

greater annual probability of flooding from the sea (>0.5%) in any year.

- *Zone 3b.* Functional floodplain – comprises land where water has to flow or be stored in times of flood, and is defined as land which would flood with an annual probability of 1 in 20 (5%) or greater in any year, or is designed to flood in an extreme (0.1%) flood, or at another probability agreed between the local planning authority and the Agency, including water conveyance routes.

Exception Test

19.   If, following the sequential test, it is not possible for the development to be located in zones of lower probability of flooding, an Exception Test can be applied. This test aims to manage flood risk while still allowing necessary development to occur. Where an Exception Test is required, decision-makers should apply it at the earliest stage possible in planning. For an Exception Test to be passed:

- it must be demonstrated that the development provides wider sustainability benefits that outweigh flood risk
- the development should be on developable previously developed land, or on a reasonable alternative site if this cannot be satisfied
- a flood risk assessment must demonstrate that the development will be safe, without increasing flood risk elsewhere, and, where possible, will reduce flood risk overall.

Responsibilities for developments in the flood plain

20.   The owner has the primary responsibility for safeguarding his land against flooding, and also managing his land drainage to prevent flooding any adjacent property. Any developer is responsible for demonstrating that the proposals are consistent with the recommendations in PPS25, providing a flood risk assessment if required, and ensuring that the development is safe from flooding and does not increase flood risk elsewhere.

21.   The local planning authority should consult the Agency and other relevant bodies when preparing policies in LDDs with respect to flood risk management. Following an amendment to Article 10 of the GDPO in 2006, local planning authorities are required to consult the Agency on all applications for development in flood risk areas (except for minor development), including flood risk from all sources, and for

any development on land exceeding 1 hectare outside flood risk areas. If the Agency objects to proposals and this objection cannot be resolved following liaison and discussion, the local planning authority may still approve a planning application. In this case, the Town and Country planning (Flooding) (England) Direction 2007 requires the local planning authority to notify the Secretary of State of the proposal, who will review the information and decide whether it would be appropriate to call it in for determination.

22.   The Agency has statutory responsibility for flood management and defence in England. It provides planning advice on flooding issues and advice on preparation of SFRAs and RFRAs. It is a statutory consultee for planning applicants as defined in the GDPO.

12

Roadside ditches

Ownership problems

1.   The question of the ownership of ditches alongside the highway and consequent maintenance responsibilities gives rise to frequent difficulties. The powers of highway authorities are constrained by statute and common law, and are narrower than generally appreciated.

2.   Where a highway authority has acquired, by agreement or compulsory purchase, the freehold of a site occupied by a road, ownership of that land will lie with the authority. However, where a highway is created by dedication, as is usually the case, ownership of the soil beneath the highway remains with the owner of the land who originally dedicated it, or his successors in title. Accordingly, the highway authority usually owns only the surface of the road and as much of the soil below, and air above, as is required for its control, protection and maintenance.

3.   The lateral extent of a highway is in general terms a question of fact, applying the presumption (subject to evidence to the contrary) that a highway extends between the hedges or fences. S.263 HA 1980 makes it clear that the whole of the 'highway', together with 'the materials and scrapings of it', vests in the highway authority. The section of the highway (vertically) that vests in the highway authority is only that which may be necessary to enable it to carry out its duty of maintenance and to enable the public to pass and to re-pass. Denning LJ said that the depth below the surface that vested in the authority may be said to be 'the top two spits' (*Tithe Redemption Commission* v. *Runcorn District Council* (1954) 2 WLR 518).

4.   Where a ditch lies between neighbouring land and the carriageway, there is a presumption that the ditch does not form part of the highway, and the highway authority therefore cannot alter it or otherwise deal with it, except with the consent of the owners. This was affirmed in *Hanscombe* v. *Bedfordshire County Council* (1938) 1 Ch. 944, where

the council claimed that it was entitled to act in relation to the ditch by common law or, if not so entitled, it was authorised to act under Section 47 of the Highways Act 1864 (now S.72 HA 1980) and Section 67 of the Highway Act 1835 (now S.100 HA 1980), but the council lost the action. Farwell J said:

> The rights of the public in a high road are to pass and re-pass along it, and their right to use the way for that purpose is not limited to that part of the way which is metalled or made up, but extends to the whole highway. When, therefore, the whole portion of a highway which is bounded by a fence or hedge is capable of being used to pass and re-pass, the whole portion is deemed to have been dedicated to the public. When, however, a portion of the whole is a ditch which *prima facie* is not adapted for the exercise by the public of their right to pass and re-pass, the presumption, in my judgement, is that the ditch does not form part of the highway. That is a presumption which may be rebutted (see *Chorley Corporation* v. *Nightingale* (1906) 2 KB 612; (1907) 2 KB 637), but the onus lies on those who assert that the ditch is part of the highway.

5.   The status of ditches in relation to the highway have been considered by the courts on many occasions, and the following decisions may be found instructive: *Chippendale* v. *Pontefract Rural District Council* (1907) 71 JP 231; *Simcox* v. *Yardley Rural District Council* (1905) 69 JP 66; and *Walmsley* v. *Featherstone Urban District Council* (1909) 73 JP 322.

6.   Highway authorities are empowered by S.100 HA 1980 to provide for the drainage of highways on land in or adjoining the highway and to take effective action if drainage works are interfered with. Compensation is payable to an owner or occupier of land who suffers damage by the exercise of these powers. A highway authority may not, in exercising these powers, interfere with a watercourse or other works vesting in a drainage authority, without the latter's consent. In *Hanscombe* v. *Bedfordshire County Council* (1938) 1 Ch. 944, a highway authority placed pipes in a ditch belonging to the owners of land abutting the highway without their knowledge or consent, and filled in the ditch. The pipes effectively drained the adjoining land and the highway. It was held that the statutory provisions did not entitle the highway authority to trespass on the ditch and to lay pipes without the permission of the owners concerned. In *Attorney General* v. *Waring* (1899) JP 789, it was held that the owner of land adjoining the highway has

a common law duty to scour and cleanse the ditches that adjoin the highway to prevent them from causing a nuisance to the highway, and that the highway authority can, notwithstanding its statutory remedies, bring an action against the owner for an injunction restraining the continuance of the nuisance.

7.   In *Provender Millers (Winchester) Ltd* v. *Southampton County Council* (1940) 1 Ch. 131, where the county council diverted the flow of a watercourse when carrying out its highway duties, it was decided that, even though the council was carrying out its duties under statutory powers, it had no general right to invade the rights of others:

> such an invasion may render them liable in damages or to an injunction, unless they can show that the work done was reasonably necessary and was properly performed in all respects, and that, if it resulted in damage, there was no way of doing it that would not have had the effect.

8.   If a highway authority has from 'time beyond memory' discharged water from the highway onto an adjoining owner's land from a highway drain, the court must presume a legal origin for this right if it is challenged and if the authority can show no document conferring the right (*Attorney General* v. *Copeland* (1902) 1 KB 690).

9.   In *King's County Council* v. *Kennedy* (1910) 2 IR 544, there was a bank by the side of the highway and a hole in the bank by means of which surface water from the highway was discharged on to the defendant's lands. There was no evidence as to the origin of the hole nor was there any defined channel on the defendant's land into which the water discharged could flow. This hole had been in existence for at least 29 years. It was held that the court ought to presume a legal origin for this outlet.

10.   In general terms, it is usual for the roadside ditches to be the responsibility of the adjoining landowner; exceptions to this rule are where the ditch was constructed to drain the highway or where it falls within land owned by the highway authority.

Filling in or culverting

11.   In a situation where a ditch appears to a highway authority to constitute a danger to users of the highway, S.101 HA 1980 gives the highway authority the power either:

(a) to fill in the ditch if it is considered unnecessary for drainage purposes, if the occupier agrees, or

(b) to place pipes in substitution for the ditch, if the occupier agrees, and to fill in the ditch (see further in Chapter 13 below).

The highway authority must pay compensation to the owner or occupier of any land who suffers damage from these actions.

12.   If the landowner wishes to carry out the piping himself, he must make allowance for the established right of discharge by the highway authority, and he will retain the responsibility for its maintenance.

13.   Before culverting any watercourse (which term includes ditches), the consent of the drainage authority is required (S.23 LDA 1991).

Powers of diversion

14.   S.110 HA 1980 empowers the highway authority to divert a non-navigable river or watercourse or to carry out any other works on any part of a watercourse including a navigable watercourse (as defined in S.111 HA 1980). The diversion may be for the purpose of highway works, the provision of new accesses, maintenance compounds, trunk road picnic areas, lorry areas and service areas. The highway authority must consult the Agency and every council in whose area the works are to be carried out, and serve notice on the owners and occupiers of land affected. Compensation can be claimed for any damage to, or depreciation of, any interest in land or interference with access to a watercourse, unless the works are carried out on land acquired compulsorily.

15.   An Order, made under S.108 HA 1980 by the highway authority and confirmed by the Secretary of State for Transport, may authorise a highway authority to divert a navigable watercourse in connection with the construction, improvement or alteration of the highway. Parts 1 and 3 of Schedule 1 to HA 1980 require copies of the order to be served on the Agency and on every navigation authority affected by the proposals.

Culverts under highways

16.   Culverts under highways are normally the responsibility of the highway authority if they were constructed to facilitate the maintenance of the highway. The culvert must be of sufficient size and depth to accommodate the normal flows from the catchment. If development takes place upstream, however, there is no statutory responsibility for the highway authority to enlarge the culvert to take the increased run-off. Accordingly, careful consideration of planning applications will be

required to avoid problems caused in this way (see generally Chapter 10 above). Similarly, if an upstream landowner wishes to drain his land with a tile drainage system and requires a lower invert level to an existing culvert, he must pay for the required work (see 13.26 below).

Acceptance of other flows

17.   Where a highway authority has used its powers to culvert a ditch or watercourse, it cannot then refuse to accept the natural run-off of surface water to it. But if, as a result of a change in conditions upstream, the system causes flooding of the highway, the authority may exercise the power of a sewerage undertaker under the WIA 1991 (see S.100 HA 1980). Before doing so, it must give notice of its intention to the relevant district council or unitary authority and the Agency within whose area the powers are proposed to be exercised.

Drainage to existing sewers

18.   S.264 HA 1980 establishes the right to drain a highway to existing drains and sewers, and the dispute resolution procedures concerning this (see also 5.48 above).

Run-off on to highways

19.   A number of engineers have reported that difficulties have arisen in dealing with the problem of the discharge of surface water from fields through farm gateways and onto the highway. While the legal position over the maintenance of roadside ditches is clear, there has been some doubt as to the powers available to deal with this problem. It has been suggested that the powers under S.163 HA 1980 could be used (see 5.47 above).

Right of statutory undertakers to discharge into a ditch or watercourse

20.   Statutory provisions relating to the pollution of water are now provided for under Part III WRA 1991. S.100 HA 1980 entitles a highway authority to discharge from a drain without committing the offence of polluting controlled waters unless the discharge contravenes a notice of prohibition under S.86 WRA 1991.

21.   At common law, a discharge may be made to a watercourse as long as the effluent does not prejudicially affect the quality of the water in the watercourse and a nuisance is not created. In *Durrant v. Branksome Urban District Council* (1897) 2 Ch. 291, it was held that a discharge could lawfully be made although it contained some sand and soil. A nuisance can be created if the discharge contains oil or other pollutants washed off the surface of the highway.

Ditches maintained by a highway authority

22.   In certain circumstances, a highway authority will take over responsibility for roadside ditches. This is regarded as a responsible working arrangement which is neither embodied in statute nor based on court decisions, as far as is known. The basis is that if work is done by the highway authority for the improvement of highway drainage, then the highway authority should be responsible for the maintenance of the new drainage works. The circumstances relevant are:

(a) where the ditch has been materially interfered with and significantly regraded by the highway authority in order to assist the drainage of the highway
(b) where the highway authority was responsible for realigning a ditch, e.g. following a highway improvement, or
(c) where the highway authority had piped a length of ditch in accordance with S.101 HA 1980.

23.   Where a highway floods as a result of an obstruction in a roadside ditch and where the landowner responsible is clearly not prepared to remove the obstruction, the highway authority will often undertake the necessary clearance work. This practice is frequently adopted for expediency, although the owner has a liability to clear the obstruction.

24.   Water passing along ditches eventually leaves the side of the highway either by being transmitted onto adjacent land or by entering a watercourse. In the case of water transmitted onto adjacent land, this can vary from what is little more than a hole in a hedge to a substantial culverted watercourse across adjacent land. The right to transmit water in this manner can be established as an easement in a formal deed, or alternatively by long usage. In either case, the highway authority would probably be liable for maintenance of the drainage facility to whatever extent is necessary to ensure the highway is effectively drained.

Highway or land drain

25.　Where water is present on a highway and is removed through a gully and highway drain, some highway authorities take the view that they are responsible for the maintenance of the drainage system only until a point is reached where water other than that originating from the highway enters the system. This principle will justify a finding that a surface water drainage system on a new housing site is a liability of the highway authority, provided that no house or land drainage is connected to it.

26.　In distinguishing between a highway drain and a drain, the courts have adopted a functional test rather than one based on the historical origins of the drain. In *Attorney General* v. *St. Ives Rural District Council* (1961) 1 All ER 265, the Court of Appeal approved the test put forward by the judge in the lower court, who said that:

> There are, of course, road drains and gullies whose main, if not whose only, function is to drain the highway. The repair of such drains and gullies is, in my judgement, clearly a function with respect to highways. There are often drains and ditches whose sole function is to drain agricultural land and which cannot in any way affect highways. The repair of such drains and ditches is equally clearly not a function with respect to highways. Then there are drains and ditches that affect both agricultural land and highways. Whether or not the function of maintaining and repairing these drains and ditches can properly be described as a function with respect to highways depends, in my view, on the degree to which the drains can be truly regarded as land drainage or road drainage.

27.　The devolution of responsibility over what are sometimes termed 'awarded drains' must therefore be investigated with care, and is likely to depend on the facts of each individual case (see also 5.52 above).

13

Culverted watercourses

Summary of problems

1. The substantial number of practical problems which arise in connection with piped and culverted watercourses generally fall into three main categories:

   (a) problems generated by engineers, riparian owners and developers, who favour culverting, without being fully aware of the practical problems and environmental disadvantages which may arise

   (b) problems associated with the cost to developers, riparian owners and/or local authorities complying with design criteria for culverts, stipulated by drainage bodies in consents required under S.23 LDA 1991, S.109 WRA 1991 or S.263 PHA 1936

   (c) problems resulting from a failure to consult with the appropriate drainage body and to obtain a written consent as above, with the result that flooding and other problems arise subsequently, mainly because of inadequate culvert design.

2. Some local authority engineers feel that this last problem is less relevant to those culverts provided for new development than it is with failures on the part of riparian owners to apply to the district council or unitary authority for permission under S.263 PHA 1936 for the culverting of ditches. It is generally agreed, though, that inadequacy of control is a common problem, and in many cases the situation is probably attributable to an ignorance of the law.

3. The Agency resists the culverting of natural watercourses for both hydraulic and environmental reasons. Its statutory responsibilities to promote and further the conservation and enhancement of watercourses and associated land (see 17.4–17.5 below) are generally in direct conflict with the principle of culverting. Developers should therefore not be surprised if a refusal to consent an application is received, or development proposals modified to incorporate open watercourses as

important landscape features to be retained and enhanced. Even where planning permission has been obtained, an enforcement notice may still be served on an illegal structure (see also 10.1 above). The Agency has published an explanatory document entitled *Environmental Policy Regarding Culverts*.

Control over culverting

4.   Statutory provisions give drainage bodies significant powers to exercise detailed control over the culverting of watercourses. In addition to their powers in connection with public health functions, local authorities, in their capacity as planning authorities, have powers which are intended to ensure that a new development is satisfactorily provided for in relation to land drainage (see Chapter 10 above).

5.   A local authority's public health powers in connection with culverting form a part of the powers which the authority may exercise under Part XI PHA 1936, in connection with watercourses, ditches, ponds, etc. (see 5.17–5.26 above).

6.   The public health powers of local authorities and the land drainage powers of drainage boards are complementary, and are exercised for different purposes. The planning powers exercised by local authorities are additional, and are more general in application. These powers constitute an important safeguard in respect of land drainage matters, which may include culverting.

7.   It will be noted that local authorities, in addition to possessing both public health and planning powers in respect of the control of culverting, also have powers to exercise such control in considerable detail. S.263 PHA 1936 states that it is unlawful to proceed with culverting work unless it is in accordance with 'plans and sections' approved by the district council or unitary authority.

8.   In considering the design of any culvert, S.263(3) PHA 1936 provides that a local authority must itself pay for any requirement imposed on the applicant regarding the provision of additional flow capacity, as compared with the quantity of water 'that he is otherwise obliged to receive or to permit to pass'. It is clear that the purpose of this section is to enable the authority to ensure, among other things, that the capacity of a proposed culvert is adequate for the future.

9.   In view of the importance of the responsibilities placed in the hands of local authorities in these matters, it is to be hoped that economic stringency will not be permitted unduly to influence a local authority's

technical decisions as to the adequacy of culvert designs submitted to it for approval.

10.   Unless a watercourse is designated as a main river, the Agency has no obligation or statutory authority to accept, as a charge on public funds, all or part of the costs of carrying out works to the technical standards which it may specify in any consent which is required for a culvert.

11.   The problems arising from failures to obtain consent under S.23 LDA 1991 or S.109 WRA 1991 are associated with the general problem of inadequate development control (described in 10.36–10.44 above).

12.   It is difficult to envisage a culverting proposal which would not be 'likely to affect the flow', and which should therefore not be referred to the drainage authority for consent. Nevertheless, this has not always been done, and it is reported that problems have arisen subsequently as a result.

13.   Even if it were contended that a consent from the Agency was not required, as there was no likelihood of affecting the flow, there would still be an additional statutory requirement to consult under S.266 PHA 1936 (see 5.25 above).

14.   A local authority may require an owner or occupier to repair, maintain and cleanse any culvert (S.264 PHA 1936) and contribute the whole or part of the cost of works required for the above purposes, or to carry out such works by agreement (S.265 PHA 1936).

Practical problems associated with culverting

15.   Developers, and some engineers, have been inclined to regard the culverting of an existing open watercourse as a simple practical solution to the problems of further development. As has been noted, however, the Agency now tends to regard the culverting of any open watercourse as environmentally undesirable, and will consent to it only if no satisfactory alternative is available.

16.   As well as the environmental undesirability of culverting open watercourses, practical advantages and disadvantages should also be borne in mind.

17.   The advantages of culverting include preventing the dumping of rubbish, reducing maintenance requirements, enabling development over a watercourse and diminishing the problems of access for maintenance. The potential problem as to who maintains the watercourse when a developer's firm goes out of existence can also be remedied by providing that subsequent responsibilities for maintenance are

transferred from the developer to another person or body by an agreement entered into at the time of culverting.

18.   To avoid the problems that arise from shared responsibility for the maintenance of a culverted watercourse, it may be adopted by a sewerage undertaker. This may be done only if the pipeline will also receive drainage from buildings and therefore fall within the definition of a public sewer (see Chapter 7 above). There may be resistance to the adoption of a culvert by a sewerage undertaker on other grounds.

19.   Usually, the flooding of a watercourse causes little, if any, permanent damage to adjoining land, especially where, historically, the hazard has been noted and provision has been made accordingly.

20.   A new piped culvert may not prevent flooding on to adjoining land when the design criteria for flow are exceeded. During severe storms, flooding upstream will occur, possibly at locations where the resulting damage may be serious.

21.   The extra costs incurred in a development project when a drainage board exercises its statutory powers to protect the public from flooding, e.g. by design requirements for culverts, are often regarded by a developer, a riparian owner, or even by a local authority, as excessive. The need for this expenditure, however, may be justified and anticipated where early consultation with the drainage board is undertaken.

22.   Once a watercourse has been culverted, it is virtually impossible for an expansion of flow capacity to be provided, except at prohibitive cost. Where enlargement is practicable, there are considerable difficulties in apportioning costs between those riparian owners who benefit, and in recovering such costs.

23.   Inadequately designed culverts can constitute a safety hazard, and instances have occurred where children have been swept into culverts and drowned. Protective grids across the opening to a culvert, designed for preventing access or blockages, will inevitably accumulate rubbish (e.g. see *Sedleigh-Denfield* v. *O'Callaghan* (1940) 1 KB 489 and *Pemberton and another* v. *Bright and another* (1960) 1 All ER 792). Even if a regular cleaning programme is provided, e.g. by arrangement with the local parish council, this programme may be insufficient to prevent flooding.

24.   A large and rapid accumulation of debris can occur within a short time, particularly at the height of a severe storm. It is highly improbable that human resources would be available immediately to clear a culvert entrance and maintain it free from obstruction throughout the duration of the storm. In such cases, flooding upstream may result, with the consequent risk of damage to property.

25.   Maintenance costs for culverts can prove to be expensive. Particular problems have been reported concerning tree roots which have blocked culverts, or the land drains leading to them. Even if the tree or trees responsible can be identified, it is sometimes difficult in practice, in view of high costs and problems of practicability, to persuade or require the owner to carry out the necessary remedial works.

Legal problems associated with culverting

26.   Engineers and planning officers who are considering the provision or approval of a culvert should be aware of the Court of Appeal decision in *Pemberton and another* v. *Bright and others* (1960) 1 All ER 792. In this case, a culvert without a grid had been installed by the highway authority to facilitate road widening. Consequently, debris blocked the entrance to the culvert and caused flooding. The court held that the ruling in *Greenock Corporation* v. *Caledonian Railway Co.* (1917) AC 556 applied, and indicated that:

> it is the duty of anyone who interferes with the course of a stream to see that the works which he substitutes for the channel provided by nature are adequate to carry off the water brought down even by extraordinary rainfall, and if damage results from the deficiency of the substitute which he has provided for the natural channel he will be liable.

27.   In accordance with this principle, various kinds of legal liability may arise from a decision to culvert a watercourse, as described in the following paragraphs (see also Chapter 7 above, which deals with some of the considerations).

28.   In law, the riparian owner will own the land over the culvert and the fabric of the pipe, and the various rights and duties of the riparian owner will continue to apply in relation to the culvert.

29.   If public funds are used, or a developer pays a contribution, in whole or in part, towards the culverting, the ownership of the pipe (and other rights and duties) becomes uncertain. In practice, it is likely that public (local authority) funds would be used at a future date for remedial works if there were to be a failure, e.g. a collapse of the culvert. Otherwise, even though the landowner may be responsible for the land and the pipe, he may not be able or willing to finance the works.

30.   Additional legal complications can arise where ownership has changed since the original provision of a culvert, and where a property search on the purchase of the property has failed to identify potential

legal liabilities of the purchaser. As a result, owners in some cases have been unaware that a culvert even exists. When they do become aware of the culvert's existence, they may be unwilling to meet their responsibilities with regard to maintenance.

31.   When an old culvert needs replacing, it can prove difficult to ascertain the extent of a riparian owner's liability in respect of the new culvert. It is unlikely that the owner will have adequate insurance cover to finance the replacement of a culvert.

32.   It may not always be practicable for a local authority to exercise its powers under S.264 PHA 1936 to require an owner to maintain, cleanse or repair the culvert. For example, this problem could arise where subsequent construction and development work, permitted by the planning authority, has made cleansing and maintenance of the culvert difficult or impossible.

33.   It is unreasonable for a local authority to serve a notice to carry out work which is impracticable. Difficulties have also arisen in exercising powers under S.259 PHA 1936, regarding watercourses which are a statutory nuisance, in cases where the riparian owner claims he did not know of the existence of the culvert. Difficulties may also be encountered in such cases in showing that the nuisance has arisen through the act, sufferance or default of the owner.

34.   A good illustration of some of the legal difficulties involved in culverting is *Dear* v. *Thames Water* (1993) WaterLaw 118. In that case, the plaintiffs house was subject to periodic flooding from a culverted watercourse, and an action in negligence and nuisance was brought both against the sewerage undertaker and the local authority which actually carried out the cleaning of the culvert. However, the action in negligence failed because, in law, the defendants only owe a duty to the public generally, rather than to individuals. The action in nuisance also failed. Although such flooding could amount to a nuisance, in this case the liability for one of the culverted streams remained with the riparian owners, which included the plaintiff, Mr Dear. Although the water company had the means to stop the nuisance under the PHA 1936, this was not enough to say that, in law, they had sufficient control over the watercourse to be successfully sued in nuisance.

35.   Given these difficulties, and others which may arise in practice, engineers and planning officers should be prepared to give careful consideration to safety, environmental, legal and technical aspects of a proposal to culvert an open watercourse. In some cases, the problems identified may be sufficient to render the proposal undesirable.

36.   It is also essential to make explicit provision for future responsibility for maintenance. This applies equally where a proposal for culverting is abandoned and an existing open watercourse retained. Some local authorities have adopted a practice of entering into agreements with developers to provide for this. In exchange for an appropriate lump sum payment from a developer, the local authority undertakes responsibility for future maintenance of the culvert or watercourse.

37.   Difficulties can arise if it is necessary for a watercourse to be culverted in order to enable a particular development to be realised. For example, even if the Agency permits culverting, it will be unlikely to relieve riparian owners of obligations to undertake future maintenance of the culvert, by the exercise of its discretionary powers over main rivers.

38.   Given the general opposition of the Agency towards culverting (see 13.3 above), it is essential for pre-planning application development enquiries to be directed to it on these matters, if delay is to be avoided. In view of the discretionary character of the powers of the Agency with regard to maintenance, responsibility for maintenance of any new culvert may remain with the owner. In this regard, the Agency is concerned about the difficulties and dangers involved with the maintenance of culverts and for these reasons may be unwilling to accept future responsibilities.

40.   Local authorities should be aware of the problems that can arise with such culverts (see also 13.15–13.25 above) and should liaise closely with the Agency on such matters.

41.   In general, the Agency is likely to avoid undertaking responsibility for the maintenance of a structure about which it has technical or other reservations, and which would otherwise be a single riparian owner's sole responsibility.

42.   In this connection, S.264 PHA 1936 provides that all owners and occupiers of land must repair, maintain and cleanse any culvert in or under that land. If it appears to the district council or unitary authority concerned that any person has failed to fulfil this obligation, then they may serve notice on that person, requiring the execution of any necessary works of repair, maintenance or cleansing.

14

Grants and contributions

Policy framework

1.   Defra encourages drainage authorities to provide, renew and improve flood defences, coast protection works, flood warning systems and related items by offering grant aid on capital expenditure. To qualify for grant aid, schemes must be technically sound, economically worthwhile and environmentally acceptable (see 2.3 above). Present policy is not to grant aid for new rural arterial drainage schemes intended primarily to increase agricultural production. References in this chapter to Defra policy and practice apply equally to the WA.

Environment Agency funding

2.   The majority of capital investment will be in grants to the Agency, who in turn fund new and improved risk management projects and defences delivered by them, by local authorities and by IDBs. The Agency also invests in reducing the consequences of flooding; through better risk mapping, strategic planning, flood forecasting and warnings, and development control.

Local authority expenditure

3.   It is for individual local authorities to decide how much to spend on flood risk management, subject to limits on overall budgets and the need for investment on other priorities. There has been an increase in spending on flood risk management since the extreme flood events in the summer of 2007.

Internal Drainage Boards

4.   IDBs are able to raise funds through charges and levies on local authorities and agricultural land owners, to fund their drainage and

risk management activity. They can also apply for capital funding for improvement works from the Agency, and receive payments from the Agency in relation to water entering IDB districts from higher ground.

Environmental aspects

5.   The WRA 1991, LDA 1991 and EA 1995 place duties on drainage authorities and Secretaries of State. Further details are given in the Code of Practice on Environmental Procedures for Flood Defence Operating Authorities (Internal Drainage Boards and Local Authorities) Approval Order 1996. Defra has also produced guidance for drainage authority managers in environmental manuals, such as *Environmental Procedures for Inland Flood Defence Works*, published in 1992, which set out recommended environmental procedures (see Chapter 17 below).

6.   The Environmental Impact Assessment (Land Drainage Improvement Works) Regulations 1999 as amended by the Environmental Impact Assessment (Land Drainage Improvement Works) (Amendment) Regulations 2005, require drainage authorities to advertise proposals for improvement works, consider representations, prepare an ES if necessary, and invite comments. Reference is made to the Secretary of State for a decision in cases of dispute. Similar provisions apply in relation to new works under the Town and Country Planning (Environmental Impact Assessment) (England and Wales) Regulations 1999, which replaced and revoked the Town and Country Planning (Assessment of Environmental Effects) Regulations 1988 (see further in Chapter 17 below). Except in very limited circumstances, Defra grant aid is not given unless these procedures are followed. Defra grant aid may be available for the preparation of an ES.

Economic assessment

7.   The costs and benefits of each potential risk management project (such as a new or improved community defence) in England are assessed by operating authorities using a standard appraisal methodology. This is based on HM Treasury's *The Green Book*, which sets common principles and approaches for all Government investment. The appraisal process results in the selection of a preferred risk management option in each case, in close consultation with interested parties including the local community.

8.   The Agency then collates information together on the costs and benefits of each potential project across England, and how each would contribute towards achieving the outcome targets set for the current spending period. In general, the Agency allocates Government funding to those projects that deliver the greatest contribution towards the outcome targets for each £1 of investment, although, in some cases, funding is provided to meet legal requirements or to fund emergency works.

9.   Other than through the setting of outcome targets, Defra ministers do not influence individual investment decisions unless the whole-life value of the project exceeds £100 million or the area strategy exceeds £250 million.

10.   *Flood and Coastal Defence Project Appraisal Guidance: Economic Appraisal* (FCDPAG3, December 1999) covers the economic aspects of project appraisal. This is one of a series of six documents (FCDPAG1–6) which is designed to provide an integrated site of guidance on all aspects of project appraisal.

Maintenance

11.   Defra requires drainage authorities to take responsibility for mainten-ance of completed works which have been grant aided.

Legislation

Water Act 2003

12.   S.147–149 WRA 1991 gave Defra the power to make grants to the Agency towards expenditure on the improvement of existing drainage works or the construction of new drainage works, flood warning schemes and flood defences. Powers exist that allow grants to assess whether drainage works should be carried out and to obtain and organise infor-mation, including information about natural processes affecting the coastline, allowing coastal defence plans to be formulated (S.101(1) EA 1995, amending S.147 WRA 1991). Plans must be approved, and the Secretary of State must be satisfied that work is properly carried out.

13.   The WA 2003 repealed S.147–149 of the Water Resources Act 1991 to enable Ministers to make block grants to the Environment Agency for flood defence works and flood warning schemes.

14.   S.69 of WA 2003 gives details of grants for drainage works and flood warning systems.

Land Drainage Act 1991

15.   S.59 LDA 1991 gives Defra the power to make grants available to IDBs and other drainage bodies for drainage schemes. These have been extended by the WA 2003. Schemes carried out by IDBs or local authorities on behalf of others and at their expense under S.20 LDA 1991, and work to repair or rebuild bridges (S.59(7) LDA 1991), can also be grant aided.

Coast Protection Act 1949

16.   Defra may also offer grant aid to maritime councils for qualifying coast protection schemes (S.21 CPA 1949). This includes grant-aiding the preparation of strategic studies. Where two-tier local government remains, the county council must make a contribution to the district council towards any expenditure incurred (S.20 CPA 1949). A draft Marine and Coastal Access Bill will simplify arrangements for managing marine developments.

Administration

Applications

17.   An application for grant aid is submitted to the Agency, and supporting documentation must include full details of the work and its cost. An engineer's report should describe the need for the work, the options considered (including doing nothing) and the reasons for selecting the preferred option. It should demonstrate the economic value of taking action and set the scheme in a strategic context. Before seeking a grant, authorities are advised to consult the Defra/WA *Flood and Coastal Defence Project Appraisal Guidance Note.* The environmental acceptability of the scheme must also be established.

18.   On receipt of the application, the regional engineer may either approve it locally, if it is a small scheme, or recommend approval by Defra headquarters.

Payments and grant rates

19.   After formal approval, authorities must adhere to the grant regulations. In particular, Defra must be notified when work starts and finishes, and final accounts must be submitted promptly. Since April 2006, Defra has grant-aided local authority capital improvement

projects at a rate of 100% of approved eligible costs. For local authority flood risk projects, the grant-aiding process is operated by the Agency.

Useful publications

20. In addition to the FCDPAG3 guidance note, practising engineers may find the following helpful in the preparation of schemes for grant aid (see the bibliography in Chapter 18 below).

- HM Treasury, *The Green Book: Appraisal and Evaluation in Central Government* (2003), which aims to make the appraisal process more consistent and transparent.
- FCDPAG3 Supplementary Note (2004), which updates the appraisal of human-related intangible impacts of flooding and allows an economic value to be included for this.
- FCDPAG3 Supplementary Note (2006) on climate change and how this applies in the decision-making process.
- Flood Hazard Research Centre at Middlesex University, *The Multicoloured Manual* (2003), which consolidates advice on calculating the benefits of flood alleviation, coastal protection and sea defence.

21. The Agency is due to publish guidance on how to apply for funding and on the new way of allocating funding in terms of contribution towards outcome measures.

Contributions

22. There are no statutory provisions requiring private sector contributions to be made to schemes when a drainage body or authority decides to exercise its drainage powers. However, authorities may decide to seek contributions.

Development

23. The Defra grant memoranda make it clear that where work will facilitate development, the planning authority is expected to seek appropriate contributions towards the cost of the scheme. These are deducted from the costs admissible for a grant.

Betterment and commutation of liabilities

24. Where a scheme results in improvement to existing structures, roads, etc., of a third party, authorities are expected to obtain contributions

towards the improvement. These are deducted from the sum eligible for grant. Payments received in commutation of liabilities are treated similarly.

Windfalls

25.　　Other contributions towards essential drainage works are considered as 'windfalls', and the grant is limited only if the total of the grant plus contribution exceeds the scheme cost.

Bridges

26.　　Some drainage boards and highway authorities have an agreed formula for calculating contributions towards the costs of bridge reconstruction. Typically, it is based on factors such as ownership, carriageway widening, waterway deepening, etc. This is an effective way of streamlining negotiations.

Flood protection grants

27.　　A £5 million flood protection grant scheme was announced in June 2009 as part of the Government's response to the Pitt Review. In the first round of funding, £3 million was made available to 25 local authorities to offer practical flood protection solutions to private properties, including air brick covers and door guards.

15

Mining subsidence and land drainage

Principal statutes

1. Land drainage can be affected by subsidence due to mineral extraction, and various powers exist to control the environmental effects of such work.

2. Coal mining is probably the most commonly occurring cause of subsidence damage, and this chapter relates largely to the effects of subsidence by coal mining. References to 'subsidence' are therefore references to 'coal-mining subsidence'.

3. The Coal Industry Act 1994 (CIA 1994) allowed for the coal-mining activities of the state-owned British Coal Corporation to be transferred to the private sector, and restructuring commenced in October 1994. A new regulatory body, the Coal Authority (CA), now owns all unworked coal and issues licences allowing operators to work coal (S.25 CIA 1994). Licensed operators therefore have operational responsibilities while the CA has a largely regulatory role.

4. The Town and Country Planning Act 1947, which came into force on the 1 July 1948, recognised mining as a form of development. Any mining which commenced before that date may be brought under some degree of planning control under the provisions of the Town and Country Planning (Minerals) Act 1981. The TCPA 1990 reaffirms that mining is a form of development. The GPDO, Schedule 2, Parts 19 and 20, designates specific aspects of mining activities by licensed operators as permitted development, but makes these subject to conditions and, in some cases, prior approval by a mineral planning authority (the county council or unitary authority) (see also 10.3).

5. The Town and Country Planning (Minerals) Act 1981, which is now incorporated into the TCPA 1990, established mineral planning authorities and gave them important powers to control the

environmental effects of mine workings by requiring planning permission for the mining and working of minerals, and 'after-care' conditions, where the land affected is subsequently to be used for agriculture, forestry or amenity purposes. These conditions include, among other things, requirements for watering or draining the land affected. The cessation of operations or the continuation subject to conditions can also be imposed on existing developments without the benefit of a permission through the making of discontinuance orders.

6.   In practice, however, control over subsidence affecting land drainage as a result of mining is limited. At present, the most important enactment is the Coal Mining Subsidence Act 1991 (CMSA 1991), although local remedies relating to watercourses are provided for under the Doncaster Area Drainage Act 1929. A number of amendments to the CMSA 1991 have been made by the CIA 1994. Where notice was served before 30 November 1991, remedies may subsist under the Coal Mining (Subsidence) Act 1957 or the Coal Industry Act 1975.

7.   In general, damage caused by subsidence must be made good, though property need not be returned to a better condition than it was before the damage occurred. Responsibility for making good such damage lies with 'the person responsible'. For working mines, this will be the licensed operator, but the CA has responsibility for the far greater number of mines that are now abandoned (S.43 CIA 1994).

Coal Mining Subsidence Act 1991

8.   The CMSA 1991, as amended, is now the principal Act applicable where damage results from subsidence arising from the lawful working of coal. The Act allows claims in respect of such damage, provided that the claimant serves a damage notice on the CA or licensee within 6 years of having the knowledge required to found a claim. This is defined as knowledge that the damage has occurred and that the nature and circumstances of the damage indicate that it may be subsidence damage. In addition, however, such knowledge includes that which might reasonably have been acquired from any observable or ascertainable facts, including those which would have been ascertainable with the help of expert advice (S.3 CMSA 1991).

9.   An alteration of the level or gradient of land or property is not considered to be subsidence damage unless the 'fitness for use' has been affected (S.1 CMSA 1991).

10.   Where the nature of the damage and the circumstances are such as to indicate that the damage may be subsidence damage, the onus is on the

CA or licensee to show that the damage is not subsidence damage (S.40 CMSA 1991). Unfortunately, as was the case under previous legislation, there is no provision for the payment of compensation for consequential loss, except in the case of 'small firms' (firms with fewer than 20 employees) (S.30 CMSA 1991). The damage, however, has to be made good to the reasonable satisfaction of the claimant (S.6 CMSA 1991). Special provisions apply to farm and crop loss payments (SS.26–28 CMSA 1991).

11.   There is a fairly elaborate system of notices and counter notices required by the Act; the major ones are as follows.

From the claimant:

- Damage Notice (see 15.8 above) (S.3 CMSA 1991 and the Coal Mining Subsidence (Notices and Claims) Regulations 1991). A Damage Notice is included in the leaflet 'Coal mining subsidence damage', available from the Subsidence Department at the CA (see 15.3 above).
- Notice of Works – minimum of 7 days (or 14 days by prior request) before remedial works are started. Where urgent works (emergency works) are required, the notice should be given as soon as reasonably practicable (SS.12 and 13 CMSA 1991 and 1991 Regulations).

From the CA or licensee:

- Notice of risk from subsidence – issued to all property and land owners. S.46(5) CMSA 1991 enables the Secretary of State for Trade and Industry to make regulations as to how and when such notices should be served, and the CA notified in turn (see the Coal Mining Subsidence (Provision of Information) Regulations 1994).
- Stop Notice – issued where the CA or licensee considers that the property is at risk from further subsidence. This notice prevents the CA or licensee from paying a double remedy, except in the case of emergency works (SS.16 and 17 CMSA 1991).
- Notice of proposed remedial action – requires the CA or licensee to notify of the intended action, where liability is accepted. A Schedule of Works should also be produced, and claimants have a short time in which to query this before it becomes legally binding (SS.4–6 CMSA 1991).

12.   The CMSA 1991 departs from the previous legislation by giving claimants the right to seek compensation for costs reasonably incurred

in repairing damage, rather than having the damage made good by what are now licensees. Additionally, claimants may request an advance payment in lieu of carrying out works. Clearly, it is for claimants to decide on the most suitable course of action. In certain cases, the CA or licensee may elect to make a depreciation payment, where the cost of remedial works would exceed the depreciation in value of the property by a prescribed margin (SS.7–15 CMSA 1991). The CA or licensee may decide that certain preventive measures may be desirable in order to minimise subsidence damage. The cost of such works will be borne by the CA or licensee, and consent cannot be unreasonably withheld (S.33 CMSA 1991).

13.   S.36 CMSA 1991 deals specifically with land drainage systems. Outside the Doncaster Drainage Area, the CA or licensee must remedy, mitigate or prevent deterioration to a land drainage system from subsidence, reasonably requested by the drainage authority. The CA or licensees may elect to make a payment equal to the cost reasonably incurred in carrying out remedial works, and to make a lump sum payment in respect of the capitalisation of any recurring costs. Regulations may be made by the Trade and Industry Secretary and Defra in respect of procedures to be followed and the determination of disputes under this section (see the Coal Mining Subsidence (Land Drainage) Regulations 1994).

14.   Various supplemental provisions in the CMSA 1991 allow for: the avoidance of double remedies (S.37); the reimbursement of successful claimants' expenses (S.38); the resolution of disputes (SS.40–42); time limits for certain disputes (S.44); and the service of documents (S.51).

Mining Codes

15.   Under what are known as the Mining Codes, notices of approach are served by the CA or licensees as a statutory duty, to safeguard certain local authorities and statutory undertakers against the consequences of subsidence. The Water Service Companies, for instance, should receive notices of approach where impending mining is likely to affect waterworks and sewage treatment works. The relevant Waterworks Code, formerly to be found in the Waterworks Clauses Act 1847, and the Water Act 1945, is now included in the WIA 1991 (Schedule 14). In addition, courtesy notices of approach are sent to some companies as well as to the Agency. These notices are a valuable source of information regarding future subsidence, and it may be possible to

reach agreement with the CA or licensee regarding the issue of courtesy notices if these have not already been issued.

16.   As far as the individual private landowner is concerned, no notices of approach are issued as a right, but anyone can approach the CA to inspect plans of past and current mine workings. All land and property owners should, in addition, receive notice of risk of subsidence under S.46 CMSA 1991.

Pollution of watercourses

17.   Coal mining may also affect land drainage through the discharge of minewater.

18.   It is the common law right of riparian owners that the natural water of a stream should flow past their land substantially unaltered in quality or quantity (see *Young & Co.* v. *Bankier Distillery Co.* (1893) AC 691; and see 6.2 above). SS.85 and 88 WRA 1991 provide that a consent is required under S.88 WRA 1991 to discharge any trade effluent, including minewater, to any controlled waters. Any discharge consent issued may include conditions covering quality, rate and volume discharged, etc. (Schedule 10 WRA 1991).

19.   S.89(3) WRA 1991 provides, though, that the offence of polluting controlled waters will not be committed by permitting water from an abandoned mine, or an abandoned part of a mine, to enter controlled waters. However, this defence does not now apply where the mine or part of a mine was abandoned after 31 December 1999 (S.60 EA 1995). Pollution permitted (but not 'caused') after this date will not therefore enjoy any special legal privilege.

20.   The EA 1995 also provides that licensees must give at least 6 months' notice of any proposed abandonment of a mine to the Agency (S.91(B) WRA 1991). Special provision is made where mines are abandoned in an emergency.

21.   By a *Memorandum of Understanding between the NRA and the Coal Authority* (1995), these organisations (and hence the Agency) have agreed procedures which apply in the event of the CA (but not the licensed operators, who are not a party to the agreement) ceasing, altering or starting to pump minewater from abandoned mineworkings.

16

Sea defence and coast protection

1.   In the past, riparian owners of coastal land were sometimes uncertain as to which authority had powers to deal with sea defence and coast protection works, and development on the coast took place with less knowledge of natural processes and risks. The DoE's *Policy Guidelines for the Coast*, was published in 1995, which also summarised Government policy in this area. This has since been superseded by Shoreline Management Plans (SMPs), which are discussed further below. Coastal erosion is now an integral part of flood risk management.

2.   In simple terms, sea defence is the alleviation of flooding of low-lying land caused by tidal flooding, for which purpose the Agency has powers under the LDA 1991 and WRA 1991. Occasionally, other organisations, including the Crown Estate, the Property Services Agency, or local authorities or IDBs, may be responsible for providing sea defences. Local authorities and IDBs have powers under the LDA 1991.

3.   An authority may seek a contribution towards the cost of works from the local authority, IDB or other beneficiary. It may also receive grants from Defra towards the cost of capital works.

4.   Coast protection is the protection of land from erosion or encroachment by the sea. This is the primary responsibility of the maritime district or unitary councils, under the provisions of the CPA 1949. However, the powers of the Agency to construct new work for flood prevention and to maintain and improve existing works extend beyond the low water mark in relation to defence against sea or tidal water, and are exercisable irrespective of whether or not they are in connection with a main river (S.165(2) WRA 1991).

5.   Within a comparatively short length, the coastline may change physically from, say, cliffs to low-lying estuary. For engineering and administrative purposes, it may therefore be sensible for such a length to be treated individually. This may be done by agreement between

the maritime council, the Agency and the government department concerned (Defra or the WA), using powers either under the CPA 1949 or the LDA 1991 and the WRA 1991.

6.   Planning Policy Guidance 20, *Coastal Planning* (PPG20), was issued in 1992, and discusses types of coasts, policies for their conservation and development, and policies covering risks of flooding, erosion and land stability, as well as coastal protection and defence. There is a 2009 consultation paper, *Development and Coastal Change*, which will replace PPG20. PPG20 sets out a planning framework for the continuing economic and social viability of coastal communities.

7.   Definitive national strategies for flood and coastal defence are contained in the Defra's Making Space for Water Strategy. The Agency has overall operational responsibility for sea and tidal defences (S.165(2) WRA 1991). It co-operates closely with local authorities, corporate bodies and private individuals, who have powers or duties in relation to the provision of sea defences. The Agency and IDBs are statutory consultees under S.5 CPA 1949 in respect of the proposed carrying out of coast protection works.

8.   Since the late 1990s, significant progress has been made in understanding and mapping coastal processes through the first generation of SMPs, which were introduced in 1995, and covered 6000 kilometres of coast in England and Wales. They provided a large scale assessment of coastal processes and risks, and a framework for reducing risks to people and the environment. Many operating authorities on a regional basis followed the recommendation and produced an SMP. Since then, there have been several major studies such as Futurecoast, Foresight, UK Climate Impacts Programme, Catchment Flood Management Plans and SFRAs.

9.   The latest *Shoreline Management Plan Guidance* was published by Defra in 2006, and gives advice on what an SMP should include and guidance on how to produce one. A second generation of Shoreline Management Plans (SMP2s) are currently in production, which aim to identify 'sustainable approach to managing coastal risks in the short, medium and long term'. SMP2s will include an action plan to prioritise specific flood and erosion risk management schemes, coastal erosion monitoring and how best to adapt to change.

10.   The Government has encouraged the formation of voluntary coastal defence groups made up of maritime district authorities and other bodies with coastal defence responsibilities such as the Agency, Natural England, port authorities and fisheries bodies. Following consultation between the groups, the Agency and RFDCs, a report was prepared

by the Agency in 2008, entitled *Coastal Groups in England*, setting out a strategic overview of sea flooding and coastal erosion risk management. The original coastal groups are to be combined to form a smaller number of larger strategic coastal groups.

11.   In 2007, new arrangements were announced for the Agency's overview role for the future management of coastal erosion and sea flooding in England. Under this strategic role the Agency will:

- take the lead for all sea flooding risk in England, and fund and oversee coastal erosion works undertaken by local authorities
- ensure that sustainable long-term SMPs are in place for the English coastline
- work with local authorities to ensure that the resulting flood and coastal erosion works are properly planned, prioritised, procured, completed and maintained to get the best value for the public purse
- ensure that third party defences are sustainable.

12.   The draft Marine and Coastal Access Bill 2008 will include a new marine planning system to enable a more strategic approach to coastal management. This Bill is due in 2009 but no dates had been published at the time of writing.

17

Land drainage and the environment

Introduction

1.   This chapter draws attention to two important features of environ-mental protection legislation affecting land drainage: the environmental duties and responsibilities of authorities concerned with drainage; and the requirement for environmental assessment of land drainage works and of development that is likely to affect drainage.

2.   The benefits of land drainage measures may often appear obvious, e.g. where they relieve urban areas or agricultural land from flooding. However, the volume and pattern of flow of surface and sub-surface water are also matters of great significance for the wider environment. Water areas, including coasts, rivers, lakes, marshland and wetlands, are vulnerable to change. Drainage proposals may affect nature and landscape conservation, water supply, recreation patterns and features of cultural value, and will have implications for general amenity.

3.   Land drainage proposals or the drainage implications of other projects can have significant adverse effects on these interests. In extreme instances, these may make the project unacceptable, despite all efforts at mitigation. In most cases, however, adverse effects can be minimised if the project is designed and implemented sensitively, and there may even be opportunities to incorporate environmental improvements.

Environmental obligations

Environment, water and land drainage Acts

4.   The environmental obligations of drainage authorities and bodies are set out in the WIA 1991 and the LDA 1991, the latter, as significantly amended by the LDA 1994, providing also for similar duties on local authorities. These duties apply also to ministers with responsibility for

106

the environment and communities. The environmental duties of the Agency are provided for in the EA 1995. Broadly, all have a general environmental duty, while exercising their statutory powers:

(a) to further the conservation and enhancement of natural beauty and the conservation of flora, fauna and geological or physiological features of special interest, so far as may be consistent with legislation relating to their functions

(b) to have regard to the desirability of protecting and conserving historic buildings, sites and objects and preserving public rights of access

(c) to take into account any effects their proposals may have on the beauty or amenity of any rural or urban area or on any such flora, fauna, buildings, sites and objects (S.3(2) WIA 1991; S.61A and 61B LDA 1991; S.7 EA 1995).

5.   The Agency also has the general duty to 'promote' conservation and enhancement of natural beauty and amenity to the extent it thinks desirable (S.6(l) EA 1995). This means effectively that the Agency not only has to be mindful of conservation while undertaking its statutory functions (which may include its functions in relation to development planning and control) but also has an overriding duty to improve the aquatic environment for conservation reasons, when considering proposals relating to its functions. The conservation duties of the Agency and the IDBs apply also when they give consent to others to carry out work (under S.61A LDA 1991).

6.   The Agency, local authorities and IDBs have a duty to consult Natural England or the CCW before authorising any works (e.g. landdrainage consents and abstraction licences) that might affect or damage a Site of Special Scientific Interest (SSSI), or land of special interest in a National Park or the Broads, that they have been notified of (S.61C LDA 1991 and S.8 EA 1995).

7.   These duties must be read alongside those contained in S.4 EA 1995 with respect to sustainable development. Thus, the principal aim of the Agency is to discharge its functions in a way which 'contributes towards attaining the objective of achieving sustainable development'. In doing this, the Agency will be guided by the advice of ministers in this objective, and must take into account likely costs. More generally, in exercising any power (but not duty), the Agency must take into account the likely costs and benefits, unless it is unreasonable to do so (S.39 EA 1995).

8.   It should be noted that the duties on local authorities and IDBs under the LDA 1991 as amended apply only in relation to their functions

under the LDA 1991. They do not seem to extend more generally to all powers exercised in relation to land drainage and flood defence matters. Thus, it seems that a local authority using powers under the PHA 1936 or HA 1980 would not be bound to act in accordance with any general environmental duties.

9. Government guidelines were published by the DoE (*Code of Practice on Conservation, Access and Recreation*, 1989) and MAFF (*Conservation Guidelines for Drainage Authorities*, 1991). Provision to make further general guidelines is contained in S.9 EA 1995. The *Conservation Guidelines for Drainage Authorities* were substantially revised and reissued in 1996 as *The Code of Practice on Environmental Procedures for Flood Defence Operating Authorities (Internal Drainage Boards and Local Authorities)* under S.61E LDA 1991. Similar provision is made for issuing conservation guidance in the Broads area under the Norfolk and Suffolk Broads Act 1988.

10. The guidelines emphasise the need to consult well in advance with Natural England (or the CCW) and other conservation bodies when drawing up maintenance programmes or planning any capital works, including changes in water resource management. It is good practice to consult statutory or voluntary organisations concerned with environmental matters on proposals that may affect their interests, even in cases where it may not be legally required.

Wildlife and Countryside Act 1981

11. The WCA 1981 originally contained some of the environmental obligations that are now enacted in the LDA 1991 and EA 1995. The WCA 1981 also gives additional protection to SSSIs. In conjunction with the more recent Acts, its provisions have the following implications besides those already described.

12. Drainage bodies and water companies may have a direct interest as owners or occupiers of land that is designated as an SSSI. In these cases, Natural England or the CCW provides them with a list of potentially damaging operations, and they must give 4 months' notice of their intention to carry out any of them. Drainage operations elsewhere in the catchment that could affect water-tables in the SSSI may be included in any management agreement.

13. A graphic illustration of the impact drainage operations may have on SSSIs can be seen in *Southern Water Authority* v. *Nature Conservancy Council* (1992) 3 All ER 481. In this case, the water authority temporarily entered land at Alverstone Marshes SSSI on the Isle of Wight to

carry out drainage works described by the House of Lords as an act of 'ecological vandalism'. However, the authority was held not to be an 'occupier' for the purposes of the WCA 1981 and was not therefore criminally liable for its actions. Given the provisions of S.4 WIA 1991 and similar provisions in relation to works carried out or authorised by public bodies on land notified to them as of special scientific interest, it must be hoped that such situations do not arise again in the future.

Conservation (Natural Habitats, etc.) Regulations 1994

4. These Regulations were made to give effect to the EU Directive on the Conservation of Natural Habitats and of Wild Fauna and Flora ('the Habitats Directive'). The Regulations provide for the designation of 'European sites' (which include sites designated under the EU Directive on the Conservation of Wild Birds (1979)) which must be protected and managed in the interests of conservation. These include marine sites. In general, the Regulations use similar mechanisms of control as contained in the WCA 1981, e.g., the notification of potentially damaging operations, but several important differences exist, not least with respect to marine sites.

5. Public bodies with powers or duties to authorise operations, or to undertake works, must exercise these after assessing the implications for European sites (Regulation 48). This includes decisions of planning authorities (Regulation 54). Where this assessment reveals that negative conservation impacts are likely, such consent or operations may only be undertaken where there is an overriding public interest. This test is particularly strong where the conservation of 'priority' habitats or species are concerned (Regulation 49). Existing consents, including planning permissions, must be reviewed where adverse conservation impacts are likely (Regulations 50, 51, 55 and 56).

6. For marine sites, any competent public authority having functions relevant to marine conservation must exercise them to secure compliance with the Habitats Directive. This includes, in particular, functions under the WRA 1991 and LDA 1991 (Regulations 3(3) and 6, 1994 Regulations). Every such authority must have regard to the requirements of the Directive in exercising their functions (Regulation 3(4)). Relevant authorities (which include local authorities, the Agency and IDBs: Regulation 5) may establish a management scheme for such sites (see also Regulation 35).

7. Where the Agency or an IDB enters into an agreement to carry out work on land in a European site, the Regulations stipulate that no

limitation on them carrying out such agreements is put on them by their constitutions (Regulation 105).

Other designations

18.   Some particularly important sites may also be designated under the Ramsar Convention on Wetlands of International Importance Especially as Waterfowl Habitat (1971). Management agreements on environmentally sensitive areas (designated under the Agriculture Act 1986) also often proscribe activities affecting water levels and drainage.

19.   Other environmental designations that may be encountered, within or near to which special care should be taken, include Special Protection Areas, Special Area of Conservation, National and Local Nature Reserves, National Parks, Areas of Outstanding Natural Beauty, Heritage Coasts (a non-statutory designation), Ancient Monuments and Listed Buildings.

20.   Other regulations, passed to give effect to EU measures incorporating environmental considerations in the Common Agriculture Policy (under EU Regulation 2078/92), also have a bearing on land drainage. In England, the Habitat (Water Fringe) Regulations 1994, Habitat (Salt-Marsh) Regulations 1994 and Habitat (Former Set-Aside Land) Regulations 1994, and, in Wales, the Habitat (Water Fringe) (Wales) Regulations 1994, all provide for payments to be made where, to varying extents, limitations on drainage activities over a period of years are agreed to. Although voluntary schemes in limited areas only, they may provide sufficient incentive to prevent damaging drainage activities being conducted, and are good illustrations of recent attempts to provide for greater proactive control in this area. For further information on, and planning policy in relation to, many of these designations, see Planning Policy Statement 9, *Biodiversity and Geological Conservation*.

21.   Within Wales, the overall land use planning framework is established by Planning Policy Wales (2002), and this is supplemented by other guidance in the form of various Technical Advice Notes (TANs) and Circulars. All of this operates within the context of the Town and Country Planning Acts.

Environmental Impact Assessment

What is Environmental Impact Assessment?

22.   EIA is the process by which information about the predicted environmental effects of a proposed project is systematically gathered, presented

110

and used as a basis for determining an application for authorisation for a development project. Used to best effect, the process identifies the need for baseline data, draws out the key environmental issues involved, promotes the generation and comparison of options, provides parameters for the design of the project, and structures the analysis and presentation of the results of the assessment so that the predicted environmental implications of the proposal are clearly stated. EIA also provides the basis for auditing and monitoring the actual environmental effects of development. The EIA process is intended to prevent or minimise adverse environmental impacts arising from potentially harmful development.

23.   An ES is the document, or series of documents, which provides information about the development and its likely environmental impact. It provides information about the development which must be taken into account by the decision-maker before a planning, or other, authorisation is granted. The ES and EIA should not be confused; EIA is a process, generally initiated by the requirements of the decision-maker, whereas an ES is the document produced by the developer to record that process, and which is submitted as part of the planning application. Since 1988, EIAs and ESs have been by law to accompany certain development proposals in accordance with the requirement upon the Government to implement EU Directive 85/337/EEC on the Assessment of the Effects of Certain Public and Private Projects on the Environment as amended by EU Directive 97/11/EC. This directive required that projects likely to have significant environmental effects should be subjected to an assessment of those effects. The amending directive came into effect in 1999, and was introduced to clarify ambiguities in the original and to and extend the scope of the projects that are subject to EIA. Not only did the number of projects subject to EIA (Annex 1) increase (e.g. wastewater treatment plants, paper and pulp production, quarrying) but it also required that the applicant be given advice on the content of the ES if requested, and that the competent authorities give its reasons for granting or refusing planning permission.

24.   Land drainage engineers will come across the legal need for EIAs and ESs in two circumstances:

(a) where an ES is required to accompany applications for consent for land drainage works
(b) where assessment of a development proposal (e.g. for an industrial estate or waste disposal) which requires an ES includes, among other things, a study of the likely effect on land drainage in the surrounding area.

25. The legal requirements for EIA are summarised in the following paragraphs. However, it is increasingly recognised that assessment of environmental effects should be informally integrated into the planning, appraisal and design of *every* relevant project (and, indeed, into the preparation of related plans, policies and programmes) if the most environmentally acceptable solution is to be achieved.

26. In planning law, environmental impact has always been a material consideration in determining planning applications, and environmental considerations must be taken into account in the drawing up of development plans. EU Directive 2001/42/EC, extending the requirement for environmental assessment to cover certain plans, policies and programmes, establishes the requirement to systematically evaluate formal development plans. Compliance with this Directive via the Environmental Assessment of Plans and Programmes Regulations 2004 has been met by the increasing use of what have become known as Strategic Environmental Assessments.

27. Additionally, EU Directive 92/43/EEC (the Habitats Directive) requires that any plan or project which is likely to have a significant effect on a designated habitats site, even though it may not be directly connected with, or necessary to, its management, should undertake an appropriate assessment of its implications for the site in the context of the site's conservation objectives. Where significant negative effects are identified, alternative options should be examined to avoid any potential damaging effects. Part IV of the Conservation (Natural Habitats, &c) Regulations 1994 implements this for specified planning and other similar consents. The Agency may, under its general powers, require some similar form of appropriate environmental assessment before applications for licences for activities (such as water abstraction, the introduction of fish, and works on the bed or banks of a watercourse), which are not covered by the formal environmental assessment requirements, are determined (e.g. see S.37 EA 1995).

When is an environmental assessment required?

28. EIA carried out in a manner that can be audited and monitored is fundamental to the formulation and planning of all projects likely to have an impact on the environment. The question 'Is a formal ES needed?' must be asked of any project that needs either planning permission or a consent or licence from a body such as Defra or the Agency. The question should be asked at the outset, and then again once it has become clear (usually at the end of the feasibility stage)

what effects the preferred option may have on the environment, and whether or not planning permission is required.

29.   New land drainage works, including flood defence works (and defences against the sea) require planning consent. Whether or not formal EIA is required for these works is determined by the Town and Country Planning (Environmental Impact Assessment) (England and Wales) Regulations 1999 (as amended by The Town and Country Planning (Environmental Impact Assessment) (Amendment) (England) Regulations 2008. There is an Explanatory Memorandum to the Regulations (DCLG, 2008) which contains an informal consolidation of the EIA Regulations as amended.

30.   Since 1994, coast protection works have been added to the list of projects (Schedule 2) for which environmental assessment of the kind required by the 1988 Regulations may be necessary where significant environmental impact is likely (Town and Country Planning (Assessment of Environmental Effects (Amendment) Regulations 1994 made under S.71A TCPA 1990; DoE Circular 7/94; and see 17.39 below).

31.   Although improvements to existing land drainage works carried out by drainage bodies and the Agency do not require planning consent, formal EIA may still be required. In these cases, reference should be made to the Environmental Impact Assessment (Land Drainage Improvement Works) Regulations 1999 as amended by the Environmental Impact Assessment (Land Drainage Improvement Works) (Amendment) Regulations 2005.

32.   The Planning and Land Drainage Regulations give legal effect in this country to the EU's Environmental Assessment Directive (see 17.25 above). A vast range of other Regulations are also in force, covering other developments that do not require planning consent but come within the terms of the Directive, such as harbour works, marine salmon farming, and electricity and pipeline works.

33.   Further information and guidance on environmental assessment can be found in DoE Circular 15/88, in DCLG Circular 02/99, which provides guidance on the Town and Country Planning (Environmental Impact Assessment) (England and Wales) Regulations 1999 for local planning authorities, and in the Detr's advisory publication *Environmental Assessment: A Guide to the Procedures* (2000). Earlier, though still useful guidance, primarily for planning officers but important also for statutory consultees and others, is given by the DoE in *Evaluation of Environmental Information for Planning Projects: A Good Practice Guide* (1994), while developers and their advisers should be aware of the DoE's *Preparation of Environmental Statements for Planning*

Projects that require Environmental Assessment: A Good Practice Guide (1995).

Where planning consent is required

34.   In the case of proposals that require planning consent, EIA is mandatory if they fall within Schedule 1 of the Town and Country Planning (Environmental Impact Assessment) (England and Wales) Regulations 1999, and may be required if listed in Schedule 2.

35.   *Schedule 1. Projects which must be assessed.* EIAs must be carried out for specified major development proposals. These are developments that are particularly intrusive environmentally, such as oil refineries, airports, thermal power stations, chemical works, inland waterways or ports, and certain types of waste incinerator.

36.   Although land drainage as such is not a Schedule 1 project, EIA of a Schedule 1 proposal may, and perhaps in the majority of cases will, include consideration of the impact that the scheme would have on drainage of the site and the surrounding area.

37.   *Schedule 2. Projects which may have to be assessed.* Schedule 2 contains a lengthy list of projects that *might* need an environmental assessment if the scheme is likely to give rise to *significant* environmental effects. The judgement of significance lies initially with the applicant (who can in any event volunteer to prepare an ES) but the planning authority may upon formal request give a preliminary ruling or 'screening opinion' as to the requirement of EIA for a development proposal (with right of appeal to the Secretary of State) or may determine that an EIA is required in relation to an application for planning permission which has actually been submitted for determination. The Secretary of State can also direct that an ES must be prepared.

38.   It is best to consult the planning authority about the need to prepare an ES for a Schedule 2 proposal at the earliest opportunity. There is guidance in Detr Circular 02/99 and in earlier Detr Circulars 15/88 and 7/94 on what constitutes significant environmental effects, e.g. development affecting a 'European site', Ramsar site or National Nature Reserve is likely to require EIA, but often a decision requires a preliminary examination of the likely impacts.

39.   The following projects in Schedule 2 are directly relevant to water management and land drainage: water management for agriculture; a salmon hatchery and an installation for the rearing of salmon; extraction of peat, coal or sand and gravel; reclamation of land from the sea; drilling for water supplies; a harbour; canalisation or flood relief

works; a dam or other installation for long-term water storage; a long-distance aqueduct; a yacht marina; a waste water treatment plant; and coast protection works.

40.   However, as with Schedule 1, the majority of other Schedule 2 projects may have implications for land drainage. The list includes the following: intensive agricultural operations; mining and quarrying; large manufacturing plant and industrial estates; housing estates and shopping centres; waste and sludge disposal; and roads.

41.   Any application for planning permission for a Schedule 2 project that the planning authority considers is likely to have significant environmental effects will not be determined unless it is accompanied by an ES.

Where planning consent is not required

42.   The Environmental Impact Assessment (Land Drainage Improvement Works) Regulations 1999 as amended by the Environmental Impact Assessment (Land Drainage Improvement Works) (Amendment) Regulations 2005 cover land drainage improvements by drainage bodies that do not require planning consent. These are developments by a drainage body, in, on or under a watercourse, or land drainage works in connection with the improvement, maintenance or repair of the watercourse or works. Operational works not above ground level are also included.

43.   Although planning consent is not required, drainage bodies are required to consider whether or not the proposed works are likely to have significant effects on the environment. If such effects are deemed likely, they must prepare an ES, inform the public, statutory consultees and interested organisations that it is available, and give them an opportunity to comment on the proposals. These procedures also apply where the minister seeks additional information from the drainage board. If, alternatively, they decide that an ES is *not* required, they must also publicise that decision.

44.   As noted in 17.34 above, similar Regulations cover a number of other activities that come within the terms of the EU Directive but do not require planning consent.

Preparing the Environmental Statement

45.   Once it has been established that environmental assessment is required, its general scope is defined by the Regulations. There is,

however, no prescribed format for the EIA or the ES. Notably, there is no obligation on the developer to consult in the preparation of the ES. Consultation is only required from the decision-making authority once the ES has been received (see 17.53–17.55 below), although it is good practice for all parties to facilitate early consultation.

46.   *Specified information.* The following information, however, must be included in an ES for it to be valid:

(a) information about the site, design, size or scale of the development

(b) data needed to identify and assess the main environmental effects

(c) a description of the likely direct and indirect effects on human beings, flora, fauna, soil, water, air, climate and the landscape; the interaction between any of the foregoing; material assets; and the cultural heritage

(d) a description of measures to avoid, reduce or remedy significant adverse effects

(e) a summary in non-technical language of the information given in items (a)–(d).

47.   *Additional information.* Additional information will often be helpful to describe the proposal, and to explain the activities involved and their impact on the environment. The Regulations encourage but do not require the provision of information on a number of matters:

(a) the physical characteristics of the proposal, and land-use requirements during construction and operation

(b) production processes, including materials used

(c) estimates of residues and emissions (including pollutants of water, air or soil, noise, vibration, light, heat and radiation)

(d) the consideration given to alternatives to the project, the processes involved and the location, and the reasons for choosing the proposal

(e) environmental effects arising from the use of natural resources, emission of pollutants and elimination of waste

(f) the methods used to forecast the effects on the environment

(g) any technical difficulties or lack of know-how encountered in compiling the specified information.

Where given, this additional information must also be summarised in non-technical language.

Best practice

48.   When preparing an ES of a drainage scheme, it is likely that there will be impacts on most of the items listed in 17.48 above. Impacts on flora, fauna, water and the landscape may be complex. They might, for example, include the following: the effects of a lowered water-table; changes to a perched water-table; exposure of peat, leading to oxidation and leaching of iron oxides; or saline intrusion. Direct and indirect effects on past and present human use of the landscape should also be considered. Archaeological sites can be damaged by lowered water-tables. Low-lying areas such as the Somerset Levels are particularly sensitive.

49.   There is no set model for an EIA, or a prescribed format for an ES. However, the EU Directive on Environmental Assessment and the related UK Regulations give guidance on what sort of development may need EIA, the processes to be followed, and the information the ES should contain as a minimum. The first stage is to identify the environmental issues, and to thereby establish whether EIA is required. This is known as 'Screening'. The second stage, once the requirement for EIA is accepted, is 'Scoping', where the issues thus identified are looked at by the parties concerned, including statutory consultees, and the scope of these issues examined. Even at this early stage it may be that detailed studies will be required to fully assess the full required scope of the assessment. The output from this Scoping stage (the Scoping Report) will set the framework for the production of the full ES, although both the ES and the Scoping Report should remain as 'live' documents: if important environmental issues come to light during the process, they must be examined. If significant, they must be included in the process and assessed. The ES is likely to contain a number of detailed sections. It may, for some large-scale or controversial developments, be a quite substantial document, but volume should not be at the expense of clarity.

50.   Criteria for judging the quality of ESs published by the Institute of Environmental Assessment have been used as the basis for the following checklist of some of the characteristics of a good ES. An ES should:

(a) Describe clearly the development, the processes and the activities involved.

(b) Describe the environment as it now is (often referred to as baseline conditions) and as it could be expected to develop if the project were not to proceed.

(c) Identify the key environmental impacts and evaluate their importance. Methodologies should be explained, and the results of consultation with expert bodies and the public outlined.

117

(d) Predict the scale of the impact, quantifying it where possible.

(e) Assess the significance of the impacts which remain after mitigation, using appropriate national and international quality standards where these exist.

(f) Estimate the types and quantities of waste and how it is to be dealt with.

(g) Explain the consideration given to alternative sites and the reasons for the final choice. Alternative processes, designs and operating conditions (which should have been considered at an early stage of project planning) should be outlined.

(h) Consider all significant impacts for mitigation and put forward measures where practicable. It should be clear how and when the measures will be carried out. Their effectiveness should be evaluated (where there is uncertainty, monitoring programmes should be proposed).

(i) Be clearly laid out with the minimum number of technical terms.

(j) Be an independent objective assessment of environmental impacts, not a best-case statement for the development (negative impacts should be given equal prominence with positive impacts).

(k) Be accompanied by a non-technical summary that must contain sufficient information to explain how the main conclusions were reached. It should briefly explain the project, the baseline conditions, proposed mitigating measures and their effectiveness, and the remaining impacts. A brief explanation of the methodology and an indication of the confidence that can be placed in the data used should also be included.

The advice in the DoE's *Good Practice Guide* remains worthwhile, and should also be consulted by those preparing ESs (see 17.35 above).

Publicity and submission procedures

51.   The Land Drainage Regulations require the drainage authority to advertise whether or not it intends to prepare an ES. If there are objections to an intention not to prepare a statement, the minister must be asked for a ruling on whether or not one has to be prepared.

52.   Under all the environmental assessment Regulations, publicity has to be given to the availability of completed ESs. The procedures vary in detail, but generally an ES has to be available to interested parties at a reasonable charge and publicised in local papers. Certain bodies, such as, in England, Natural England and, in Wales, the CCW, must

be consulted as a matter of course. Unless the developer has formally undertaken to carry out these consultations, they remain the responsibility of the planning authority. These pre-decision consultations are in addition to those bodies which must be consulted under Article 10 of the GDPO if the development requires planning permission.

53.   The legal requirements represent a minimum level of consultation. For major or controversial projects, in particular, consultations with expert bodies and the public should feature as early as possible in the development of the project as an integral component of the ES. Many applicants may prefer to adopt a more open approach towards publicity and consultation as a matter of good practice rather than in consequence of any legal obligation to do so.

18

Bibliography

This bibliography includes references both to works referred to in the preceding text, as well as other works of use and interest.

Books, updated looseleaf volumes, reports and articles

Bailey S.H. *Garner's Law of Sewers and Drains*, 9th edition. Shaw, Crayford, 2007.

Bates J.H. *Water and Drainage Law.* Sweet & Maxwell, London (two releases per year).

Bell S. (ed.) *Water Law.* Wiley, Chichester (a journal dealing with all aspects of the law relating to water management).

CIRIA. *Sustainable Urban Drainage Systems (SUDS) manual.* CIRIA, London, 2007.

CIRIA. *Scope for Control of Urban Runoff*, Vols 1 and 2. Reports Nos. 123 and 124. CIRIA, London, 1991.

CIRIA. *Control of Pollution from Highway Drainage Discharges.* Report 142. CIRIA, London, 1994 (also constitutes NRA R&D Report 16).

Cross C.A. *Encyclopaedia of Environmental Health Law and Practice.* Sweet & Maxwell, London (three releases per year).

Cross C.A. and Garner J.F. *Encyclopaedia of Highway Law and Practice.* Sweet & Maxwell, London (three releases per year).

Department of Environment, Food and Rural Affairs. *Flood and Coastal Defence Project Appraisal Guidance: Economic Appraisal* (FCDPAG3). HMSO, London, 2000.

Department of Environment, Food and Rural Affairs. *Flood and Coastal Defence Project Appraisal Guidance: Approaches to Risk* (FCDPAG4). HMSO, London, 2000.

Department of Environment, Food and Rural Affairs. *Flood and Coastal Defence Project Appraisal Guidance: Environmental Appraisal* (FCDPAG5). HMSO, London, 2000.

Department of Environment, Food and Rural Affairs. *Flood and Coastal Defence Project Appraisal Guidance: Strategic Planning and Appraisal* (FCDPAG2). HMSO, London, 2001.

Department of Environment, Food and Rural Affairs. *Management Statement.* HMSO, London, 2002.

Department of Environment, Food and Rural Affairs. *Making Space for Water.* HMSO, London, 2005.

Department of Environment, Food and Rural Affairs. *Shoreline Management Plan Guidance*. HMSO, London, 2006.

Department of Environment, Food and Rural Affairs. *Coastal Groups Report*. HMSO, London, 2007.

Department of Environment, Food and Rural Affairs. *Coastal Groups in England*. HMSO, London, 2008.

Department of Environment, Food and Rural Affairs. *Towards a new National Flood Emergency Framework*. HMSO, London, 2008.

Department of Environment, Food and Rural Affairs. *Flood and Coastal Defence Project Appraisal Guidance: Overview* (FCDPAG1). HMSO, London, 2009.

Department of Environment, Food and Rural Affairs. *Appraisal of Flood and Coastal Erosion Risk Management*. HMSO, London, 2009.

Department of Environment, Food and Rural Affairs. *Surface Water Management Plan Technical Guidance*. HMSO, London, 2009.

Department of the Environment. *Policy Appraisal and the Environment*. HMSO, London, 1991.

Department of the Environment. *Environmental Appraisal of Development Plans*. HMSO, London, 1993.

Department of the Environment. *Evaluation of Environmental Information for Planning Projects: A Good Practice Guide*. HMSO, London, 1994.

Department of the Environment. *Preparation of Environmental Statements for Planning Projects that require Environmental Assessments: A Good Practice Guide*. HMSO, London, 1995.

Department of the Environment. *Water Act 1989: Code of Practice on Conservation, Access and Recreation*. DoE, London, 1989.

Environment Agency. *Living on the Edge. A Guide to the Rights and Responsibilities of Riverside Occupation*. HMSO, London, 2007.

Garner J.F. and Bailey S.J. *The Law of Sewers and Drains*, 8th edn. Shaw, Crayford, 1995.

Gardiner J.L. Environmentally sensitive river engineering: examples from the Thames catchment. *Journal of Regulated Rivers Research and Management*, 1988, 2(3), 445–469.

Gardiner J.L. *River Projects and Conservation: A Manual for Holistic Appraisal*. Wiley, Chichester, 1991.

Gaunt J. (ed.) *Gale* on *the Law on Easements*, 18th edn. Sweet & Maxwell, London, 2008.

Hawke N. and Parpworth N. *Encyclopedia of Environmental Health Law and Practice*. Sweet & Maxwell, London (updated regularly).

Heneage A. *Land Drainage in England and Wales*. HMSO, London, 1951.

Holmes N., Ward D. and Jose P. *The New Rivers and Wildlife Handbook*. The Royal Society for the Protection of Birds, Horsham, 1994.

Howarth W. *Wisdom's Law of Watercourses*, 5th edn. Shaw, Crayford, 1992.

Howarth W. and Brierley A. *Infiltration Drainage – Legal Aspects*. CIRIA Project Report No. 25 (also constitutes NRA R&D Report 488). CIRIA, London, 1995.

Hydro Research and Development Ltd. *Urban Drainage: The Natural Way*. Conflo Committee, Hydro Research and Development Ltd, Clevedon.

Institution of Municipal Engineers. *Seminar on Land Drainage – Whose Responsibility?* Institution of Municipal Engineers, London, 1980.

Institution of Municipal Engineers. *Symposium on Land Drainage*. Institution of Municipal Engineers, South West District, London, 1982.

Ministry of Agriculture, Fisheries and Food. *The View of the Land Drainage Powers of Water Authorities and Local Authorities*. Consultation Paper. MAFF, London, 1978.

N'Jai A., Tapsell S.M., Taylor D., Thompson P.M. and Witts R.C. *FLAIR 1990 – Flood Loss Assessment Information Report*. Flood Hazard Research Centre, London, 1988.

Orlik M. *An Introduction to Highway Law*, 3rd edition. Shaw, Crayford, 2007.

Parker D.J., Green C.H. and Thompson P.M. *Urban Flood Protection Benefits: A Project Appraisal Guide*. Gower Technical Press, London, 1987 (the 'Red Manual').

Penning-Rowsell E.C., Green C.H., Thompson P.M., Coker A.M., Tunstall S.M., Richards C. and Parker D.J. *The Economics of Coastal Management*. Belhaven Press, London, 1992 (the 'Yellow Manual').

Pitt M. *Learning Lessons from the 2007 Floods*. HMSO, London, 2008.

Rowsell E.C. and Chatterton J.B. *The Benefits of Flood Alleviation: A Manual of Assessment Techniques*. Saxon House, Farnborough, 1978.

Royal Society for the Protection of Birds/National Rivers Authority Wildlife Trust. *The New Rivers and Wildlife Handbook*. RSPB, Sandy, 1994.

Severn Trent Water Authority. *A Unified Approach to Land Drainage*. Severn Trent Water Authority, Birmingham, 1977.

Suleman M.S., N'Jai A., Green C.H. and Penning-Rowsell E.C. *Potential Flood Drainage Data: A Major Update*. Flood Hazard Research Centre, London, 1988.

Waverley J.A. *Report of the Departmental Committee on Coastal Flooding*. HMSO, London, 1954.

Wilkins J.L. Land drainage legislation and the engineer – a review and discussion, Parts I and II. *Ch. Mun. E*, 1980, 107, May–June.

Appendix 1

Relevant statutes, Statutory Instruments, EU legislation and current Government policy guidance to which reference is made

Statutes

Highways Act 1835
Waterworks Clauses Act 1847
Highways Act 1864
Public Health Act 1875
Rivers Pollution Prevention Act 1876
Public Health Acts Amendment Act 1890
Law of Property Act 1925
Doncaster Area Drainage Act 1929
Land Drainage Act 1930
Public Health Act 1936
Town and Country Planning Act 1947
Coast Protection Act 1949
Coal Mining (Subsidence) Act 1957
Town and Country Planning Act 1971
Local Government Act 1972
Coal Industry Act 1975
Reservoirs Act 1975
Land Drainage Act 1976
Highways Act 1980
Town and Country Planning (Minerals) Act 1981
Wildlife and Countryside Act 1981
Food and Environmental Protection Act 1985
Local Government Act 1985
Agriculture Act 1986
Local Government Finance Act 1988
Norfolk and Suffolk Broads Act 1988

Environmental Protection Act 1990
Town and Country Planning Act 1990
Coal Mining Subsidence Act 1991
Land Drainage Act 1991
Planning and Compensation Act 1991
Water Industry Act 1991
Water Resources Act 1991
Local Government Act 1992
Coal Industry Act 1994
Land Drainage Act 1994
Local Government (Wales) Act 1994
Environment Act 1995
Water Act 2003
Civil Contingencies Act 2004
Planning and Compulsory Purchase Act 2004
Government of Wales Act 2006
Marine and Coastal Access Bill (Draft) 2008–2009
Flood and Water Management Bill (Draft) 2009

Statutory instruments

Town and Country Planning (Assessment of Environmental Effects) Regulations 1988 (SI 1988, No. 1199)

Land Drainage Improvement Works (Assessment of Environmental Effects) Regulations 1988 (SI 1988, No. 1217)

Coal Mining Subsidence (Notices and Claims) Regulations 1991 (SI 1991, No. 2509)

Town and Country Planning (Development Plan) Regulations 1991 (SI 1991, No. 2794)

Waste Management Licensing Regulations 1994 (SI 1994, No. 1056)

Habitat (Water Fringe) Regulations 1994 (SI 1994, No. 1291)

Habitat (Former Set-Aside Land) Regulations 1994 (SI 1994, No. 1292)

Habitat (Salt-Marsh) Regulations 1994 (SI 1994, No. 1293)

Coal Mining Subsidence (Provision of Information) Regulations 1994 (SI 1994, No. 2565)

Conservation (Natural Habitats, etc) Regulations 1994 (SI 1994, No. 2716)

Coal Mining Subsidence (Land Drainage) Regulations 1994 (SI 1994, No. 3064)

Habitat (Water Fringe) (Wales) Regulations 1994 (SI 1994, No. 3100)

Town and Country Planning (General Permitted Development) Order 1995 (SI 1995, No. 418)

Town and Country Planning (General Development Procedure) Order 1995 (SI 1995, No. 419)

Town and Country Planning (General Development Procedure) Order 1995 Land Drainage Improvement Works (Assessment of Environmental Effects) (Amendment) Regulations 1988 (SI 1988, No. 2195)

The Code of Practice of Environmental Procedures for Flood Defence Operating Authorities (Internal Drainage Boards and Local Authorities) Approval Order 1996

EU Directives and Regulations

Council Directive 79/409/EEC on the Conservation of Wild Birds

Council Directive 85/337/EEC on the Assessment of the Environmental Effects of Certain Public and Private Projects on the Environment

Council Directive 92/43/EEC on the Conservation of Natural Habitats and of Wild Flora and Fauna Council Regulation 2078/92/EEC on Agricultural Production Methods Compatible with the Requirements of the Protection of the Environment and the Maintenance of the Countryside

Council Directive 2001/42/EC on the Assessment of the Effects of Certain Plans and Programmes on the Environment

Environmental Assessment of Plans and Programmes Regulations 2004

Environmental Impact Assessment (Land Drainage Improvement Works) Regulations 1999 (SI 1999, No. 1783) as amended by the Environmental Impact Assessment (Land Drainage Improvement Works) (Amendment) Regulations 2005

Council Directive 2007/60/EC on the Assessment and Management of Flood Risks

Government policy guidance

Circulars

DoE 15/88, Environmental Assessment (WO 23/88), Amendment of Regulations (WO 20/94)

DoE 11/94, Environmental Protection Act 1990: Part II. Waste Management Licensing

General Development Order Consolidation 1995, Circular 9/95, 11/95

The Use of Conditions in Planning Permissions, Circular 11/95

The Standard National Requirements for validation of Planning Applications under CLG, Circular 02/2008

DoE circular 02/99 which provides guidance on the Town and Country Planning (Environmental Impact Assessment) (England and Wales) Regulations 1999

Planning Policy Guidance Notes

PPG13, *Transport*
PPG14, *Development on Unstable Land*
PPG20, *Coastal Planning* (DoE/WO, 1992)

Planning Policy Statements

PPS1, *Delivering Sustainable Development*
PPS9, *Biodiversity and Geological Conservation*

PPS12, *Local Spatial Planning*
PPS25, *Development and Flood Risk*

Code of Practice

The Code of Practice on Environmental Procedures for Flood Defence Operating
Authorities (Internal Drainage Boards and Local Authorities) 1996

Appendix 2

List of cases to which reference is made

Attorney-General v. Copeland (1902) 1 KB 690
Attorney-General v. Peacock (1926) 1 Ch. 241
Attorney-General v. St. Ives Rural District Council (1961) 1 All ER 265
Attorney-General v. Waring (1899) JP 789
Bickett v. Morris (1866) 30 JP 532
Blount v. Layard (1891) 2 Ch. 681
Box v. Jubb (1879) 3 Ex. D. 76
British Railways Board v. Tonbridge and Malling District Council (1981) 79 LGR 565
Cambridge Water Co. v. Eastern Counties Leather (1994) 1 All ER 53
Chasemore v. Richards (1859) 7 HL Cas. 349
Chippendale v. Pontefract Rural District Council (1907) 71 JP. 231
Chorley Corporation v. Nightingale (1906) 2 KB 612; (1907) 2 KB 637
Clark v. Epsom Rural District Council (1929) 1 Ch. 287
Dear v. Thames Water (1993) Water Law 118
Durrant v. Branksome Urban District Council (1897) 2 Ch. 291
East Suffolk Catchment Board v. Kent (1940) 4 All ER 527
Falconar v. South Shields Corporation (1895) II TLR 223
Fisher v. Winch (1939) 2 All ER 144
George Wimpey and Co. Ltd v. Secretary of State for the Environment (1978) JPL 776
Gibbons v. Lanfestey (1915) 84 LJPC. 158
Grampian Regional Council v. City of Aberdeen District Council (1984) 47 P&CR 633
Greenock Corporation v. Caledonian Railway Co. (1917) AC 556
Hanbury v. Jenkins (1901) 2 Ch. 401
Hanscombe v. Bedfordshire County Council (1938) 1 Ch. 944
Harrison v. Great Northern Railway Co. (1864) 3 H&C 231
Home Brewery plc v. Davis and Co. (1987) 1 All ER. 637
Hudson v. Tabor (1877) 2 QBD 290
Hunter and others v. Canary Wharf Ltd (1995) The Times, 13 October
Hutton v. Esher Urban District Council (1974) Ch. 167
Leakey v. National Trust (1980) QB 485
King's County Council v. Kennedy (1910) 2 IR 544
Mason v. Shrewsbury and Hereford Railway Co. (1871) 25 LT 239
Meader v. West Cowes Local Board (1892) 3 Ch. 18
Menzies v. Breadalbane (1828) 3 Bli. NS 414

Nicholls v. Marsland (1875) 33 LT 265
Nield v. London & North Western Railway Co. (1874) LR 10 Ex. 4
Pemberton and another v. Bright and others (1960) 1 All ER 792
Provender Millers (Winchester) Ltd v. Southampton County Council (1940) 1 Ch. 131
Pyx Granite Co. Ltd v. Ministry of Housing and Local Government (1958) 1 QB 554
R. v. Dovermoss Ltd (1995) Env. LR 258
R. v. Southern Canada Power Co. (1937) 3 All ER 923
Rouse v. Gravelworks Ltd (1940) 1 KB 489
Rylands v. Fletcher (1868) 19 LT 220
Sedleigh-Denfield v. O'Callaghan (1940) 3 All ER 349
Simcox v. Yardley Rural District Council (1905) 69 JP 66
Smith v. River Douglas Catchment Board (1949) 113 J. 388
Southern Water Authority v. Nature Conservancy Council (1992) 3 All ER 481
Stephens v. Anglian Water (1987) 3 All ER 379
Stollmeyer v. Trinidad Lake Petroleum Co. (1918) AC 485
Taylor v. St. Helen's Corporation (1877) 6 Ch. D. 264
Tesco Stores Ltd v. Secretary of State for the Environment (1995) 2 All ER 636
Thomas v. Gower Rural District Council (1922) 2 KB 76
Tithe Redemption Commission v. Runcorn District Council (1954) 2 WLR 518
Walmsley v. Featherstone Urban District Council (1909) 73 JP 322
Wheeler v. Saunders (1995) 2 All ER 697
Young & Co. v. Bankier Distillery Co. (1893) AC 691

Appendix 3

Useful addresses

Department of Food, Environment and Rural Affairs

Customer Contact Unit
Eastbury House
30–34 Albert Embankment
London SE1 7TL

The postal address for contacting all Defra's ministers is:

Defra
Nobel House
17 Smith Square
London SW1P 3JR

Emails can be sent to ministers via the Defra Helpline:

helpline@defra.gsi.gov.uk.

Environment Agency

National Customer Contact Centre
PO Box
544 Rotherham S60 1BY
Tel.: 08708 506 506

Communities and Local Government

Eland House
Bressenden Place
London SW1E 5DU

Riverwalk House
157–161 Millbank
London SW1P 4RR

Ashdown House
123 Victoria Street
London SW1E 6DE

Scottish Office

The Office of the First Minister
St. Andrew's House
Regent Road
Edinburgh EH1 3DG

Office of the Permanent Secretary
Sir John Elvidge
St. Andrew's House
Regent Road
Edinburgh EH1 3DG

Director General Environment
Richard Wakeford
The Scottish Government
Victoria Quay
Edinburgh EH6 6QQ

Legal and Parliamentary Services
25 Chambers Street
Edinburgh EH1 1LA

Welsh Office

The Wales Office
Gwydyr House
Whitehall
London SW1A 2NP

Coal Authority

Subsidence Department
200 Lichfield Lane
Mansfield
Nottinghamshire NG18 4RG
Tel.: 01623 427162

Mining Records
Bretby Business Park
Ashby Road
Burton on Trent
Staffordshire DEI5 0QD
Tel.: 01283 553463

Index